Wireless Networks

Series Editor

Xuemin Sherman Shen, University of Waterloo, Waterloo, Canada

The purpose of Springer's Wireless Networks book series is to establish the state of the art and set the course for future research and development in wireless communication networks. The scope of this series includes not only all aspects of wireless networks (including cellular networks, WiFi, sensor networks, and vehicular networks), but related areas such as cloud computing and big data. The series serves as a central source of references for wireless networks research and development. It aims to publish thorough and cohesive overviews on specific topics in wireless networks, as well as works that are larger in scope than survey articles and that contain more detailed background information. The series also provides coverage of advanced and timely topics worthy of monographs, contributed volumes, textbooks and handbooks.

Quan Shi • Yinxin Bao • Qinqin Shen •
Zhenquan Shi • Ruifeng Gao

Connected Vehicles Traffic Prediction

Emerging GNN Methods

Springer

Quan Shi
School of Transportation and Civil Engineering
Nantong University
Nantong City, Jiangsu, China

Yinxin Bao
School of Transportation and Civil Engineering
Nantong University
Nantong City, Jiangsu, China

Qinqin Shen
School of Transportation and Civil Engineering
Nantong University
Nantong City, Jiangsu, China

Zhenquan Shi
School of Transportation and Civil Engineering
Nantong University
Nantong City, Jiangsu, China

Ruifeng Gao
School of Transportation and Civil Engineering
Nantong University
Nantong City, Jiangsu, China

ISSN 2366-1186 ISSN 2366-1445 (electronic)
Wireless Networks
ISBN 978-3-031-84547-5 ISBN 978-3-031-84548-2 (eBook)
https://doi.org/10.1007/978-3-031-84548-2

This Springer imprint is published by the registered company Springer Nature Switzerland AG
The registered company address is: Gewerbestrasse 11, 6330 Cham, Switzerland

Acknowledgments

This work was supported by the National Natural Science Foundation of China (62476145); the Humanity and Social Science Foundation of the Ministry of Education of China (24YJAZH126); the 6th "333 Talents" Technology Research and Development Talent Foundation of Jiangsu Province; and the Transportation Technology and Achievement Transformation Foundation of Jiangsu Province (2024G01).

Contents

Chapter 1
Introduction

1.1 Artificial Intelligence in Connected Vehicles

The rapid advancement of connected vehicle (CV) technologies is reshaping the future of transportation. Connected vehicles, which communicate with each other (V2V), infrastructure (V2I), pedestrians (V2P), and networks (V2N), represent a transformative step toward safer, more efficient, and more intelligent transportation systems. By enabling real-time data exchange among vehicles and infrastructure, connected vehicle technology enhances traffic management, improves road safety, and paves the way for autonomous driving capabilities. This interconnected ecosystem enables vehicles to make informed, coordinated decisions, reducing traffic congestion, minimizing accidents, and optimizing fuel consumption, thus contributing to a more sustainable transportation environment. In China, the development of intelligent transportation systems (ITS) has accelerated, with substantial advancements in traffic data collection, storage, and analysis technologies. These advancements have propelled ITS to new heights, creating unprecedented opportunities for efficient traffic management, precise resource allocation, and enhanced transportation services. One of the core components of ITS is traffic prediction, which serves as the foundation for effective traffic planning, management, control, and dissemination of information. Accurate traffic prediction provides critical decision support for traffic managers and aids drivers in route planning, reducing congestion and enhancing travel efficiency. Figure 1.1 illustrates the current situation of urban road congestion.

With the rapid development of intelligent transportation systems in China, there has been significant progress in traffic data collection and storage technologies. These advancements have created opportunities for efficient traffic management, accurate allocation of traffic resources, and high-quality transportation services. Traffic prediction plays a crucial role in the development of intelligent transportation systems as it supports traffic planning, management, control, and information

Q. Shi et al., *Connected Vehicles Traffic Prediction*, Wireless Networks,
https://doi.org/10.1007/978-3-031-84548-2_1

Fig. 1.1 Urban road congestion

dissemination. The goal of traffic prediction is to accurately and efficiently forecast the future conditions of the transportation system, providing guidance to traffic participants. This includes providing decision-making support for traffic managers and helping travelers plan their routes and allocate traffic resources effectively to mitigate or avoid congestion. However, the complex topology and dynamic characteristics of traffic networks have made traffic data voluminous and structurally complex. Successfully modeling spatiotemporal data generation and capturing spatiotemporal correlations have been ongoing challenges in achieving accurate and fine-grained traffic prediction.

The role of smart transportation is to improve the efficiency, safety, and sustainability of urban transportation systems. It utilizes advanced technologies and data analysis to optimize traffic flow and provide intelligent transportation services. Accurate traffic flow prediction is crucial for smart transportation.

Smart transportation plays a vital role in managing urban traffic by utilizing various technologies such as intelligent traffic signals, adaptive traffic management systems, and real-time traffic monitoring. These systems help to alleviate congestion, reduce travel time, and enhance the overall efficiency of the transportation network.

Accurate traffic flow prediction is essential for smart transportation. By analyzing historical traffic data, monitoring real-time traffic conditions, and employing predictive modeling techniques, it becomes possible to anticipate traffic patterns, identify potential bottlenecks, and optimize traffic management strategies. This enables transportation authorities and drivers to make informed decisions regarding route planning, mode selection, and travel time optimization.

The significance of accurate traffic flow prediction lies in its ability to mitigate congestion, improve traffic management, and enhance overall transportation experiences. By accurately predicting traffic patterns and providing real-time information, smart transportation systems can help residents and commuters make better-informed choices regarding their travel routes, modes of transportation, and departure times. This not only improves efficiency but also reduces fuel consumption, lowers emission levels, and contributes to a more sustainable and environmentally friendly transportation ecosystem.

In summary, smart transportation and accurate traffic flow prediction are instrumental in improving urban transportation efficiency, reducing congestion, enhancing safety, and promoting sustainable transportation practices. By leveraging advanced technologies and data-driven insights, smart transportation systems provide a pathway toward a more efficient, reliable, and convenient transportation network. Figure 1.2 shows the intelligent transportation system.

The study and development of neural networks have witnessed significant milestones and breakthroughs, leading to profound implications across a spectrum of domains, including healthcare, finance, transportation, and entertainment. The interdisciplinary nature of neural network research has drawn expertise from diverse fields such as computer science, mathematics, neuroscience, and engineering, fostering an integrative approach that advances the understanding and application of neural network models.

One hallmark achievement in the evolution of neural networks is the advent of deep learning, a subfield of machine learning that has propelled the capabilities of artificial intelligence to unprecedented levels. Deep learning architectures, comprising multiple layers of interconnected neurons, have demonstrated remarkable prowess in processing vast quantities of data, uncovering intricate patterns, and making high-level abstractions. This capacity has catalyzed transformative breakthroughs in fields such as computer vision, natural language processing, and speech recognition.

In the realm of computer vision, convolutional neural networks (CNNs) have emerged as a dominant force, enabling machines to perceive and comprehend visual information with remarkable accuracy [1]. Applications of CNNs span diverse domains, including facial recognition, object detection, medical imaging analysis, and autonomous driving systems. The ability of CNNs to extract hierarchical features from raw pixel data has revolutionized the analysis and interpretation of visual content, driving advancements in areas like image classification, semantic segmentation, and image generation.

In the domain of natural language processing (NLP), recurrent neural networks (RNNs), long short-term memory networks (LSTMs), and transformer architectures

Fig. 1.2 Intelligent transportation system

have revolutionized the processing and understanding of textual data. These neural network models have powered a new wave of language-based applications, ranging from machine translation and sentiment analysis to chatbots and intelligent assistants. The fusion of neural networks with language models and attention mechanisms has underpinned significant strides in language understanding and generation, paving the way for more human-like interactions with AI systems.

Moreover, the integration of reinforcement learning with neural networks has unlocked new frontiers in autonomous decision-making and control. Reinforcement learning algorithms, combined with deep neural network policies, have propelled the development of autonomous agents capable of mastering complex tasks, including game playing, robotic manipulation, and resource management. The synergy between reinforcement learning and neural networks has sparked innovations in areas such as self-driving vehicles, industrial automation, and adaptive resource allocation, laying the groundwork for intelligent systems capable of learning and adapting to dynamic environments.

The advancing frontiers of neural networks also intersect with ethical considerations and societal impact. As AI systems powered by neural networks assume roles of increasing significance in decision-making processes, the need to address issues of fairness, transparency, and accountability becomes paramount. Research endeavors have thus aimed at enhancing the interpretability and fairness of neural network models, fostering a responsible and ethical deployment of AI technologies.

Looking ahead, ongoing efforts in neural network research are focused on addressing key challenges, such as robustness to adversarial attacks, efficient training and inference procedures, and the ethical implications of AI deployment. Collaborative initiatives, spanning academia, industry, and regulatory bodies aim to shape the trajectory of neural network research, ensuring that AI technologies grounded in neural networks contribute positively to societal progress and human well-being.

In conclusion, the evolution of neural networks stands as a testament to humanity's relentless pursuit of understanding and emulating the complexities of the human brain. From the early conceptualizations of artificial neurons to the sophisticated deep learning architectures of today, neural networks have redefined the landscape of artificial intelligence, driving innovations that have transcended traditional boundaries and empowered transformative applications across diverse domains. As the journey of neural network research continues, the prospects for pioneering advancements and interdisciplinary collaborations herald a future where intelligent systems, underpinned by neural networks, enrich human endeavors and redefine the possibilities of technological innovation. Figure 1.3b shows the structure of an artificial neural network.

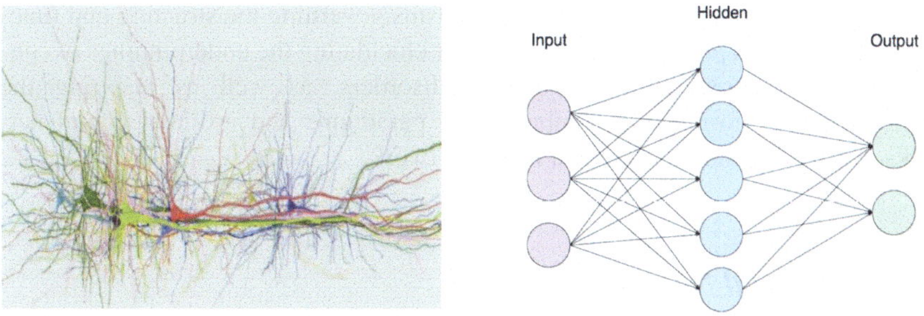

(a) The main structure of neurons (b) Neural network structure diagram

Fig. 1.3 Structure diagram of artificial neural network. (**a**) The main structure of neurons. (**b**) Neural network structure diagram

1.1.1 Biological Neural Network

Biological neural networks (BNNs) serve as the foundational architecture underpinning the cognitive and behavioral functions of living organisms. They composed of an intricate web of neurons, cells, and synapses and form the biological substrate of consciousness and cognition. Within the human nervous system, the staggering complexity of BNNs is underscored by the presence of approximately 86 billion neurons, each intricately interconnected through thousands of synapses.

At the core of BNNs lie neurons, the fundamental units responsible for information processing and propagation within the network. A typical neuron exhibits a complex structure, comprising a cell body and cellular processes. The cell body, or soma, serves as the central hub housing the nucleus and various receptors, including ion channels and neurotransmitter receptors. These receptors play a pivotal role in modulating the intra- and extracellular potential, thereby shaping the neuron's physiological response, whether through excitation or inhibition.

The cellular processes extending from the cell body, namely dendrites and axons, facilitate the transmission of signals within the neural network. Dendrites function as the receptive surfaces of the neuron, receiving sensory inputs and conveying them toward the cell body. Their branching structure enables the integration of diverse synaptic inputs, thereby influencing the neuron's overall excitatory or inhibitory state. Conversely, axons serve as the conduits for transmitting the neuron's output to other neurons or effector tissues. Each neuron typically possesses a singular axon, which can branch into collaterals to establish connections with a multitude of target cells, thus propagating neural signals across the network.

This intricate interplay of neurons, synapses, and cellular processes endows BNNs with remarkable computational capabilities, enabling the cognitive and sensorimotor functions essential for the survival and adaptation of living organisms. The dynamic modulation of synaptic strength, the spatial and temporal integration of synaptic inputs, and the propagation of action potentials collectively underpin the complex information processing and emergent behaviors exhibited by biological neural networks. Understanding the mechanisms governing the structure and function of BNNs holds profound implications for elucidating the underpinnings of consciousness, cognition, and neurological disorders, as well as inspiring the development of bio-inspired computational paradigms and artificial neural networks. Figure 1.4 shows biological neural network.

1.1.2 Artificial Neural Network

Artificial neural networks (ANNs) represent a versatile class of computational models that draw inspiration from the remarkable information processing capabilities of biological neural networks. They comprise interconnected nodes or "artificial neurons" (ANNs) that emulate the parallel distributed processing observed in their biological counterparts. The fundamental building blocks of artificial neural networks, artificial neurons, collectively form layers within the network, with each neuron establishing connections to numerous others, thus enabling the propagation and transformation of input information.

A typical artificial neuron encompasses several key components, including input weights, an activation function, and an output. The input weights modulate the strength of connections to other neurons, thereby influencing the impact of incoming signals. The activation function processes the aggregated input signals and determines the neuron's activation state, thereby governing the transmission of processed information to subsequent neurons within the network. This collective interplay of input weights, activation functions, and output functionality enables artificial neurons to capture and manipulate complex patterns and relationships within the input data.

One of the defining attributes of artificial neural networks is their capacity for learning and adaptation, a process commonly achieved through training. During training, the network systematically adjusts its internal parameters in response to input data, with the overarching objective of improving its performance on specific

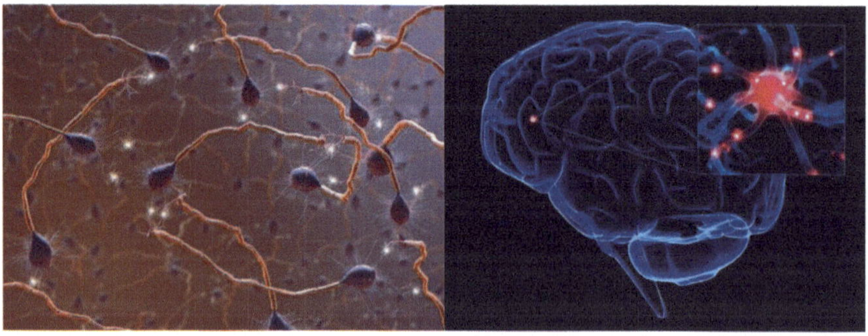

Fig. 1.4 Biological neural network

tasks. This process mirrors the process of synaptic plasticity observed in biological neural networks, wherein the strength of synaptic connections is dynamically modulated in response to experience and environmental stimuli.

The wide-ranging applicability of artificial neural networks is underscored by their efficacy in diverse domains, including machine learning, pattern recognition, and decision-making. Through their unparalleled capacity for capturing intricate patterns and extracting meaningful representations from complex datasets, ANNs have emerged as indispensable tools for addressing a spectrum of real-world challenges. From image and speech recognition to natural language processing and predictive modeling, artificial neural networks have demonstrated their proficiency in discerning intricate patterns, effectively recapitulating the information processing and decision-making prowess exhibited by biological neural networks.

By harnessing the principles underpinning biological neural networks, artificial neural networks have positioned themselves as pivotal assets in advancing technological frontiers, offering unprecedented avenues for enhancing computational intelligence and enabling transformative advancements across a spectrum of interdisciplinary domains. Figure 1.5 shows the artificial neural networks.

In the past decade, research on artificial neural networks has made significant strides, leading to breakthroughs in various fields such as pattern recognition, intelligent robotics, automatic control, prediction and estimation, as well as applications in biology, medicine, and economics. Artificial neural networks have successfully addressed many practical problems that are challenging for conventional computers, demonstrating strong intelligent capabilities.

The development of artificial neural networks has undergone several phases, including the nascent stage, the first wave, the second wave, and the third wave. During these phases, advancements in network architectures, training algorithms, and the scalability of neural networks have significantly expanded their potential applications and improved their overall performance.

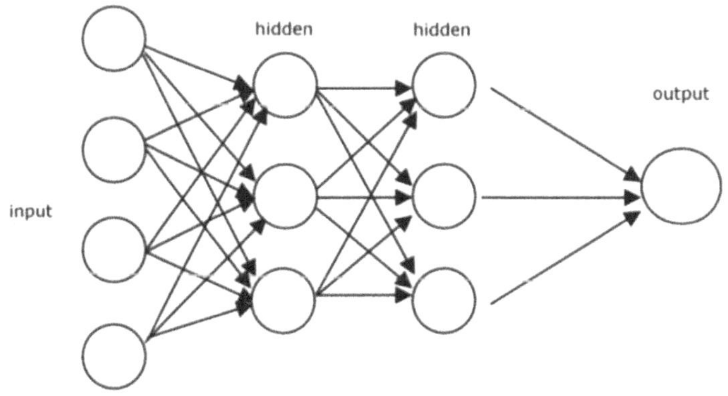

Fig. 1.5 Artificial neural networks

1.1.2.1 The First Wave (Starting in the 1960s)

The first wave of artificial neural network development emerged in the 1960s. The groundwork for this phase was laid in 1943 when American neurophysiologist Warren McCulloch and mathematician Walter Pitts introduced a formal neuron model that mimicked the behavior of biological neurons. Their model, known as the M-P [2] model, used a physical network constructed from resistors and other components to simulate neural activity.

In 1958, psychologist Frank Rosenblatt proposed the perceptron, a type of neural network that could adjust its connection weights after training, enabling a computer to make decisions based on these weights. This innovation sparked renewed interest and significant progress in neural network research, marking the first peak in its development.

However, in 1969, Marvin Minsky and Seymour Papert pointed out the limitations of perceptrons in solving linear indivisibility problems. This discovery led to a downturn in neural network research as it highlighted the fundamental challenges and limitations of the early neural network models. Despite this setback, it set the stage for the evolution of more advanced neural network architectures and training algorithms in the subsequent phases of development. Figure 1.6 shows Hebb.

1.1.2.2 The Second Wave (Starting in the 1980s)

In 1986, researchers including David Rumelhart, Geoffrey Hinton, and Ronald Williams proposed the Back Propagation (BP) algorithm, a significant breakthrough that addressed the issue of linear inseparability by introducing multiple layers of perceptrons. This development paved the way for the resurgence of neural networks, marking the second peak in their development.

Following the introduction of the backpropagation algorithm, in 1989, Yann LeCun and others successfully implemented LeNet-5, a 7-layer convolutional neural network, achieving an impressive 98% accuracy in recognizing handwritten digits. This milestone further fueled the momentum of neural network research,

Fig. 1.6 Hebb

attracting a surge of interest and contributions from numerous scholars and researchers. Additionally, significant work was conducted around the artificial neural network method proposed by John Hopfield, contributing to a research boom in artificial neural networks starting from the mid-1980s.

However, this period of advancement was relatively short lived. Toward the end of the twentieth century, the popularity of support vector machines (SVM) and other machine learning algorithms began to rise. Importantly, these other machine learning algorithms surpassed neural networks in terms of implementation efficiency. Additionally, the emergence of new algorithms such as PageRank for the Internet redirected scholars' focus toward alternative machine learning techniques, causing a shift in research away from neural networks and leading to a second decline in their popularity. Figure 1.7 shows David Rumelhart.

1.1.2.3 The Third Wave (Starting from 2000s)

During the third peak of deep learning, the resurgence of neural networks was facilitated by the confluence of several key technological and environmental factors. Advancements in hardware technology played a pivotal role, with high-speed GPU parallel computing greatly accelerating the training of deep networks with numerous layers, reducing training times from weeks or months to mere days. This computational performance enhancement proved crucial in driving the widespread adoption of deep learning across diverse domains, enabling rapid iteration and refinement of complex neural network architectures.

Simultaneously, the widespread availability of the internet and the digitalization of various industries facilitated the collection of vast amounts of training data. The availability of extensive and diverse datasets served as a critical factor in the success of deep learning models, effectively suppressing overfitting and enhancing their applicability in real-world scenarios.

Fig. 1.7 David Rumelhart

Fig. 1.8 Hinton

The open-source movement played a crucial role in democratizing access to deep learning frameworks and tools. The development and widespread availability of deep learning open-source frameworks, such as TensorFlow, PyTorch, and Keras, empowered researchers, developers, and practitioners to leverage state-of-the-art neural network architectures and algorithms for diverse applications including image and speech recognition, natural language processing, autonomous systems, and more.

As a result of these advancements and favorable external conditions, the widespread adoption of deep learning led to transformative applications in fields such as healthcare, finance, automotive technology, and beyond. From medical image analysis to autonomous driving systems, deep learning became a cornerstone of innovation, driving progress, and advancements across a wide spectrum of industries.

In conclusion, during the third peak of deep learning, the resurgence of neural networks, rebranded as deep learning, was propelled by a confluence of factors including hardware advancements, dataset availability, open-source collaboration, and the pioneering work of researchers such as Geoffrey Hinton [3]. This period marked a turning point in the history of artificial intelligence and set the stage for the transformative impact of deep learning on technological advancements and real-world applications [4, 5]. Figure 1.8 shows Hinton.

1.2 Fundamental Knowledge

1.2.1 Traffic Map

Utilizing network structure in graph-based traffic prediction enables more accurate modeling of spatial dependencies. Traffic prediction tasks heavily rely on capturing spatial correlations, which are often enhanced by incorporating rich attribute information. Traditionally represented as grid-based images, traffic data has now undergone a structural transformation into graph representations, allowing for more flexible and effective modeling of complex road networks. Early research was to integrate the road, collect network information and traffic status, obtain the spatio-temporal matrix, and then convert it to replace with images and introduce CNN models to capture traffic spatiotemporal images and then predict based on the characteristics of. There have also been studies that have improved it, constructing a two-dimensional convolutional neural network model by combining spatial attention modules, and capture the inbound passenger flow characteristics of other stations during the pre-event period. The above method for extracting spatiotemporal features is essentially based on Euclidean methods spatial data, while transportation networks have significant graph features, such as nodes have distinct local structures, and the interactions between nodes are dynamic State change. GNN can be obtained through message passing between nodes' spatial information, where the adjacency matrix is a microcosm of the graph structure and is an aggregation. The key to integrating valuable feature information tailored to different application scenarios. Different types of adjacency matrices can be constructed for graph-based traffic prediction, which are mainly categorized into static, state-dependent, adaptive, and spatiotemporal matrices, each capturing different aspects of spatial and temporal dependencies.

Static matrices are constructed assuming that the correlation between nodes does not vary with time. Various adjacency matrices have been designed by researchers to capture the diverse predefined relationships between nodes in a transportation network, including connectivity, distance proximity, functional similarity, and time series correlation.

Connectivity matrix: The simplest connectivity matrix is defined as a binary graph, where 1 indicates a connection between nodes and 0 indicates no connection. This can be expanded to describe the connectivity between regions that are geographically distant but can be efficiently reached through highways or rail transportation. Additionally, connectivity matrices can also be constructed based on station passenger flow transfer matrices and free-flow reachability matrices.

Distance matrix: The proximity between nodes can be measured based on the geometric distance between them. Specifically, the distance measure can be calculated using a threshold Gaussian kernel function to determine the proximity between node pairs. Non-neighboring nodes can be measured in terms of the number of intermediate edges or through a weighted combination of the two.

The functional similarity matrix measures whether nodes have similar traffic patterns based on the assumption that nodes with similar functions tend to have similar traffic modes or characteristics, such as the commuting traffic pattern between residential areas and office areas during rush hours, or the traffic features of the central business district during workdays and weekends. The similarity in node function is often measured by calculating the distribution of Point of Interest (POI), which is used to compute the functional similarity matrix or regional type division.

The historical mode correlation matrix measures the correlation of historical traffic patterns between nodes, taking into consideration the semantic connection between historical data. Even though nodes may not be directly connected in space, their traffic history data may contain implicit correlations that can be used as a semantic connection of historical data for measuring the correlation between nodes. The distance between historical time series data of nodes can be calculated using the dynamic time warping (DTW) method to describe the correlation of historical patterns.

These methods measure and analyze the traffic patterns among nodes to better understand and predict traffic flow and travel behavior.

Adaptive matrix: To model the dynamic correlation between nodes more accurately, an adaptive matrix can be constructed to capture the spatial relationship between nodes from time series data. The static matrices mentioned earlier, whether based on network structure (adjacency matrix) or calculated through data (semantic matrix), fail to reflect the dynamic nature of spatial correlation over time. However, traffic networks exhibit significant dynamism, and the traffic network characteristics can vary at each timestamp. To address this, an adaptive matrix is proposed. It can uncover the inherent connections between nodes through operations like node embedding and parameterized learning, resulting in an adaptive matrix that captures the potential spatial correlation and dynamic relationship patterns in traffic prediction tasks.

Space-time matrix: To capture both spatial and temporal correlations concurrently, the aforementioned methods of constructing matrices are carried out independently at each timestamp to capture the connectivity of the traffic network at a specific instance. However, for complex space-time networks like traffic networks, spatial and temporal neighbors are equally important. The features of each node are influenced not only by its neighboring nodes at the same timestamp but also by adjacent neighbor nodes from different timestamps. To encode complex space-time relationships, the introduction of spatiotemporal edges can form a space-time graph to capture the more intricate space–time relationships in traffic prediction tasks. Figure 1.9 shows graph.

1.2.2 Traffic DATA

With the improvement of intelligent transportation infrastructure, the collection of traffic data has become more convenient. To develop higher level intelligent transportation, data sharing and circulation are indispensable. The data sources used for traffic prediction mainly include sensor data deployed on the road network, GPS trajectory data, ride-hailing/bike-sharing order data, AFC swipe card data, road attribute data, POI data, and weather data. According to the temporal and spatial characteristics of the dataset, traffic data can be classified into temporal-static data, time-dynamic data, and spatiotemporal-dynamic data. Moreover, according to the data structure, it can be divided into point-based data and network-based data.

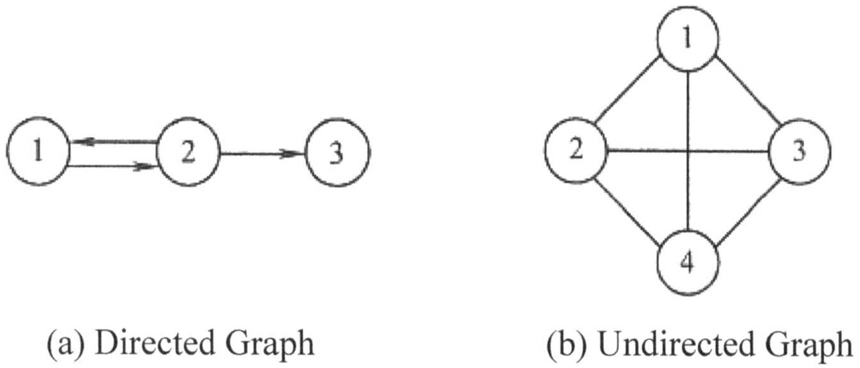

Fig. 1.9 Graph. (**a**) Directed graph. (**b**) Undirected graph

Network-based data is commonly used for traffic prediction tasks. The diversified data for traffic prediction is beneficial. For example, incorporating social media data into traffic prediction can help capture rare abnormal passenger flow characteristics. Various classification methods may be used to classify traffic data, geographic data, and environmental data. Figure 1.10 shows spatiotemporal correlation of traffic data.

Table 1.1 lists the basic information of a series of publicly available datasets, including the number of nodes and edges, the granularity of statistics, and the total time steps. The listed information represents the results after sampling. For example, the PeMSD4 dataset contains data collected from 3848 sensors on 29 roads, and after sampling, information from 307 nodes is retained. The BEIJING-subway dataset aggregates subway card swiping data at different time granularities, specifically 10-, 15-, and 30-min intervals.

The METR-LA dataset and the Performance Measurement System-BAY (PeMS-BAY) dataset are provided by the California Department of Transportation. The METR-LA dataset consists of traffic speed statistics collected from 207 sensors deployed on Los Angeles highways during the period from March 1, 2012, to June 30, 2012. Additionally, the PeMS dataset comprises real-time data collected from over 39,000 sensors deployed in the primary metropolitan areas of California's highway network. Various sub-dataset versions exist, differing in terms of time and space coverage, information type, and the number of sensors included. The PeMS

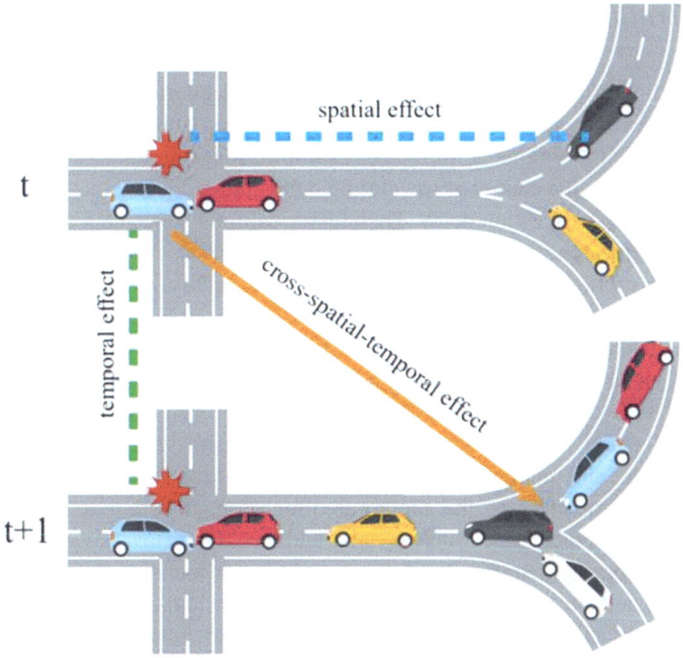

Fig. 1.10 Spatiotemporal correlation of traffic data

Table 1.1 Basic information about public datasets

Type	Dataset	Node	Edge	Statistical granularity/min	Total time steps
Sensor	METR-LA	207	1515	5	34,272
	PEMS-BAY	325	2369	5	52,116
	PEMSD3	358	547	5	26,208
	PEMSD4	307	340	5	16,922
	PEMSD7	883	866	5	28,224
	PEMSD7(M)	228	\	5	12,672
	PEMSD8	170	295	5	17,856
Taxi trajectory	SZ-taxi	156	532	15	2976
Subway card swipe	SHMETRO	288	958	15	6072
	HZMETRO	80	248	15	1650
	BEIJING-subway	276	630	10/15/30	2700/1800/900

BAY dataset, for instance, contains traffic speed statistics collected from 325 sensors in the Bay Area during the period from January 1, 2017, to June 30, 2017.

High-quality publicly available datasets are conducive to conducting comprehensive comparative experiments to evaluate the performance of models.

1.2.3 Traffic Prediction Task

Traffic prediction tasks involve various application scenarios. Each task has common characteristics such as nonlinearity, complex spatiotemporal patterns, and randomness, as well as unique features. For example, congestion delay prediction first requires analyzing the causes of travel time delay and modeling methods for calculating travel time congestion delay. Then historical time series data related to delay calculation attributes is used for prediction. Pedestrian trajectory prediction requires modeling of motion interactions and trends. Additionally, some studies have utilized multi-task learning frameworks.

Traffic state prediction involves forecasting the future state of a specific traffic object at a given time. The traffic object can be a sensor, an intersection, a road segment, or a subway station, while the traffic state can refer to traffic flow, speed, queue length, and more. There is a certain relationship among flow, speed, and occupancy rate, and incorporating multiple feature attributes as inputs can enhance prediction performance. In the temporal dimension, different feature attributes exhibit similar characteristics such as daily and weekly cycles, as well as trends. In the spatial dimension, traffic conditions at different locations have varying and dynamic impacts on the same location and are also influenced by traffic signals in urban road scenarios. Moreover, there are similar interdependencies between public transportation networks and road networks. Therefore, a single model can be applicable to multiple traffic prediction tasks. Efficient traffic state prediction enables more accurate matching of supply and demand, facilitating fine-grained traffic management such as travel route planning and congestion mitigation. Figure 1.11 shows relationships between traffic flow density and speed and prediction results.

Travel demand prediction: Predicting the number of transportation services needed in a certain area in the future to better allocate resources, such as predicting the demand for subway passengers and shared bicycles. It can be divided into region-based or station-based demand prediction and travel OD demand prediction, the latter of which can provide richer travel activity information and be more conducive to resource allocation.

Journey time and congestion prediction: Journey time is an important indicator of traffic congestion, and journey time prediction refers to inferring the time to reach the destination with a given trajectory. This problem needs to comprehensively consider the dynamic characteristics (traffic flow, speed, etc.) and static characteristics (road width, length, etc.) of the road network and aggregate dynamic and static characteristics under network conditions to infer journey time. Congestion prediction can identify potential congestion nodes in the network or predict congestion delay time. Due to the large number of commuters during rush hours, road congestion and full public transportation cause travel speed to decrease and waiting time to extend, resulting in low traffic network efficiency. To allow travelers and managers to grasp the future operation of the network in advance, feasible solutions are adopted for journey time and congestion predictions.

Human/vehicle trajectory prediction: The trajectory prediction task aims to predict the future trajectory of the target (such as pedestrians, vehicles, and other traffic participants) based on the current or historical trajectory and environmental information. Trajectory prediction results are one of the important information for

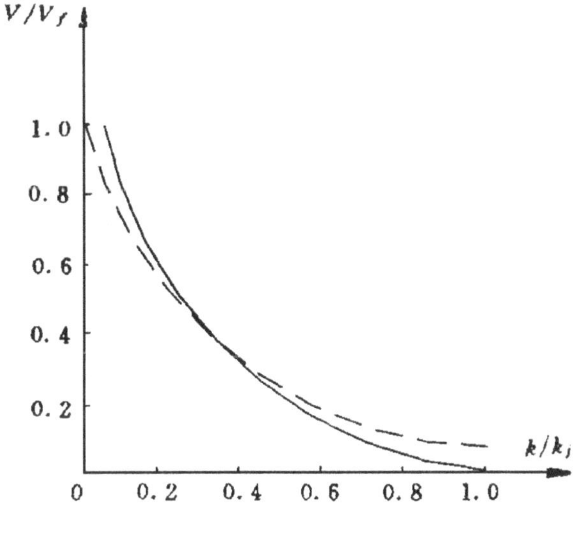

Fig. 1.11 Relationships between traffic flow density and speed

autonomous driving systems to make early decisions. How to output safe and reliable prediction results under the constraints of traffic rules is crucial for autonomous driving systems.

1.3 Connected Traffic Flow Prediction-Related Work

Traffic prediction has undergone a transformation from model driven to data-driven, while data-driven has evolved from statistical methods to traditional machine learning algorithms and then to deep learning algorithms.

With the advancement of data collection and storage technology, more and more research focuses on data-driven approaches. Early data-driven methods relied on statistical methods, inferring the relationships between different variables in the data in order to characterize temporal patterns. Representative models include linear regression models [6], integrating moving average autoregressive models [7], Kalman filtering model [8], etc. Statistical methods assume that time series come from the history and current situation of random variables and have continuity. However, due to the complex characteristics of traffic data such as periodicity and impulsivity, these methods are limited by their ability to capture nonlinear relationships. Traditional machine learning methods can fully capture traffic time ordinal numbers.

To model nonlinear relationships in traffic prediction, various machine learning methods have been applied, such as Artificial Neural Networks (ANN) [9], Support Vector Machines (SVM) [10], and k-Nearest Neighbors (KNN) [11], as discussed in the work by Vanajakshi and Rilett [12]. A comparison was made between SVM and ANN, and it was proven that both models have good dynamic response performance. In the case of missing data samples, SVM performs better. These methods have shown good performance in solving small samples, high-dimensional data, and nonlinearity, but they are limited by shallow architectures and manual feature selection methods. With the development of deep learning, end-to-end deep belief networks (DBNs) [13], (Convolutional Neural Networks, CNN) [14], (Recurrent Neural Networks, RNN), (Long Short-Term Memory, LSTM), (Gate Recurrent Unit, GRU), etc. have emerged [15–17]. These methods stand out in temporal prediction tasks and are widely used in traffic prediction scenarios. They focus on capturing temporal correlations and effectively reveal the time-varying characteristics of traffic data [18]. By capturing features across different time slices, the horizontal periodicity and vertical randomness of traffic sequences can be effectively modeled [19]. In addition, long- and short-term time series can be merged into a unified feature set to enhance long-term feature representation. By designing experiments to determine the optimal time delay, the predictive performance can be further improved [20]. These methods enrich the modeling of temporal correlations, but overlook the complex spatial correlations within the topological connections of transportation networks. In order to simultaneously capture temporal and spatial correlations, many scholars combine these basic models into different architectures [21]. However, the connectivity of the transportation network is not a two-dimensional grid, and its nodes are discrete and irregularly arranged, with asymmetric characteristics. The relationships between nodes are easily affected by distance, so CNN cannot accurately capture the spatial characteristics of the transportation network.

The transportation network is a typical graph structure with characteristics such as disorder, irregularity, and dynamism. In the real world, a large amount of data can be represented as graph relational structures containing rich information. Researchers have developed a strong interest in the extension of deep learning methods in graphs. By drawing on the ideas of convolutional neural networks, recurrent neural networks, and deep auto-encoders, a neural network structure was designed for processing graph data, and GNN emerged [22]. GNN obtains dependency relationships in the graph through information transmission between nodes in the network. In addition to strong correlation relationships that directly affect each other (such as upstream and downstream relationships), there are also implicit spatial correlations based on dynamic travel patterns of travelers (such as travel paths) among transportation network nodes. This demonstrates the high adaptability of the GNN model to transportation network data structures. GNN is widely used in the field of traffic prediction [23–26], it has played a powerful role in the field of traffic prediction. Figures 1.12 show common traffic prediction methods.

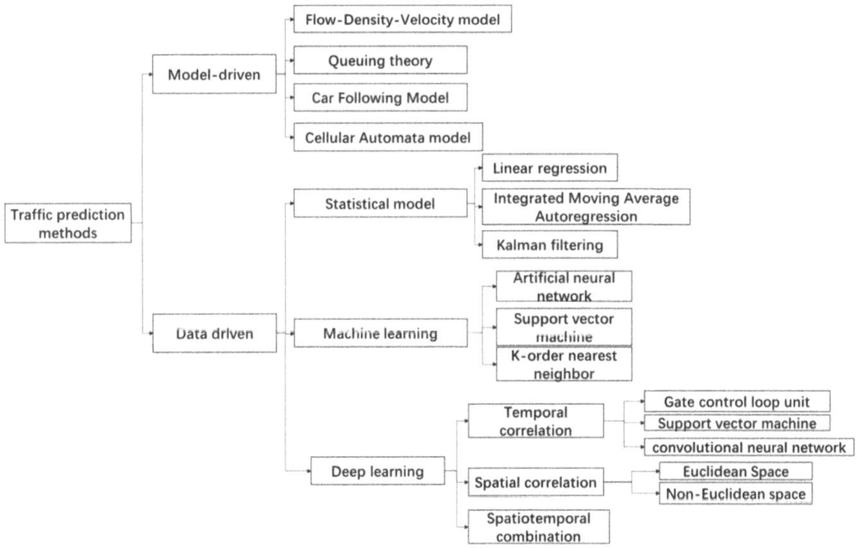

Fig. 1.12 Common traffic forecasting methods

1.4 The Development History of Graph Neural Networks

With the development of spatiotemporal data modeling techniques, the methods for traffic prediction, as one of the typical tasks in spatiotemporal computing, are continually evolving. The transportation system exhibits dynamic and complex spatial correlations, and graph data can encode diverse relationships. Graph neural networks (GNNs) demonstrate a high adaptability to transportation networks. From the perspective of graph construction, there has been a progression from predefined graphs to mining the semantic relevance in data, and ultimately to dynamically generating graphs based on the data. These developments provide solutions for capturing global and local spatial correlations.

Furthermore, traffic prediction is a time series forecasting task, and it exhibits multi-scale periodicity in the time dimension. Both local and global time correlations are equally important. Current solutions integrate multiple time slices or design multi-resolution convolutional modules to address this issue. The challenge lies in effectively integrating spatiotemporal modules to capture diverse spatiotemporal correlations and achieve real-time accurate prediction performance for various prediction tasks.

Taking these factors into consideration, this paper briefly reviews the development of traffic prediction and categorizes graph construction, data types, and application scenarios. It primarily focuses on GNN models applied to traffic prediction problems. The paper analyzes the spatiotemporal modules, graph construction factors, and modeling factors covered by GNN models for different traffic prediction tasks and examines the impact of different components on prediction performance. The paper also explores research challenges and opportunities from the perspectives of multi-source data, diverse applications, multimodality, dynamics, model interpretability, uncertainty, and small-sample learning.

Traffic networks are complex systems with dynamic behaviors at both spatial and temporal scales. Modeling the dynamic nature of traffic networks is crucial for capturing spatiotemporal features. Graph spatiotemporal networks provide a powerful framework for modeling such dynamics by considering both the dynamic inputs of nodes and the dependencies between nodes.

In the context of traffic prediction, there are two main categories of graph spatiotemporal networks: those based on recurrent neural networks (RNNs) and those based on convolutional neural networks (CNNs). RNN-based methods have been used to address the challenges of gradient vanishing and exploding commonly encountered in traditional RNNs. Researchers have often combined graph convolutions with RNN variants to overcome these issues and effectively capture spatiotemporal correlations in traffic networks.

Graph neural networks (GNNs) are versatile models capable of capturing dependency relationships between objects, and they have found widespread applications in various domains such as natural language processing, computer vision, and recommendation systems. In the field of traffic prediction, GNNs have also been extensively used. Commonly used GNN architectures for traffic prediction include convolutional GNNs (ConvGNN), graph attention networks (GAT), graph autoencoders (GAE), and graph spatial-temporal networks (GSTN).

In recent years, there has been increasing research demonstrating the power of graph neural networks (GNNs) in modeling spatiotemporal dynamics in traffic networks, facilitating tasks such as traffic prediction and congestion control. In 2005, Marco Gori et al. introduced the concept of GNNs [6] in their paper, which allowed for direct learning on graph data without losing structural information as was the case with traditional methods of transforming graphs into vectors during pre-processing.

GNNs extend the structure of neural networks by utilizing graph convolution operations. Unlike in traditional neural networks where each neuron is tightly connected, in GNNs, each neuron is only connected to adjacent neurons to simulate node relationships in a graph. This approach allows the network to retain the structure of the graph, thus addressing issues faced by traditional neural networks in processing graph data.

Through the use of GNNs, traffic congestion, traffic flow, and other factors related to traffic networks can be modeled and predicted. For instance, road networks can be represented as directed graphs, where each edge represents a road between two intersections and each node represents an intersection. By using traffic flow, vehicle speeds, and other indicators to describe the states of nodes and edges on this directed graph, traffic prediction and congestion control can be achieved.

GNNs have drawn significant attention in the field of traffic modeling. For example, some studies have utilized GNNs to predict traffic flow at intersections, thereby improving traffic control efficiency. Researchers have also used GNNs to address traffic signal control problems to alleviate congestion. Overall, GNNs have emerged as an important tool in traffic modeling and prediction, improving the accuracy and efficiency of traffic control.

In 2009, Dr. Franco Scarselli expanded upon the concept of graph neural networks (GNNs) [7] with some key improvements that built upon earlier research. Specifically, Scarselli and his team developed an algorithm that utilized GNNs to learn node representations in graph data. However, earlier forms of GNNs relied on iterative processes that utilized recurrent neural networks (RNNs) to propagate information from neighboring nodes until a fixed state was reached. While effective, this approach consumed a significant amount of computational resources.

Scarselli's algorithm, which he referred to as the Recursive Neural Network (RecNN), sought to overcome these limitations by utilizing a sequence of neural network layers that process the graph data in a bottom-up fashion. Specifically, the algorithm operates by recursively applying a set of functions to each node's feature vector, which incorporates information from the node's immediate neighbors. This process continues until the whole graph is processed and produces a fixed-size representation, called a graph embedding, that encodes the hierarchical structure of the graph in a continuous space.

The RecNN model also helps to alleviate some of the over-smoothing issues that were observed in earlier forms of GNNs by adding random noise to the node embeddings during each iteration. This encourages the model to learn more distinct representations for each node, which in turn helps it to differentiate between the nodes better.

In 2012, convolutional neural networks (CNNs) [8] demonstrated remarkable success in computer vision tasks and researchers began exploring their application to graph neural networks (GNNs). In 2013, a student of Yann LeCun, Joan Bruna, proposed the Graph Convolutional Network (GCN) model, which introduced the concept of frequency domain convolution for graph data. This approach revolutionized the field by introducing techniques to adapt convolutional operations to graph structures.

The GCN model operates by performing convolutional operations directly on the graph structure, allowing for the extraction of local features and hierarchical representations from graph data. Instead of relying solely on node features, as in traditional neural networks, GCNs leverage both node features and graph connectivity to capture relational information between nodes in the graph. By incorporating information from neighboring nodes, the GCN model can learn meaningful representations that encompass both the local and global structure of the graph.

One key advantage of GCNs is their ability to handle graphs of arbitrary size and structure. Unlike traditional convolutional neural networks that operate on fixed-size grids, GCNs can adapt to graphs with varying numbers of nodes and irregular connectivity patterns. This flexibility enables GCNs to address a wide range of graph-related tasks, including node classification, link prediction, and graph classification.

However, one limitation of the original GCN model is the high computational complexity associated with frequency domain convolution. The model requires processing the entire graph at once during computation, which results in expensive matrix decompositions and increases the time complexity of the operation. This can be particularly challenging when working with large-scale graphs.

In 2016, Kipf et al. made a crucial breakthrough in the field of graph convolution by introducing a simplified definition of frequency domain convolution. This innovation enabled graph convolutional operations to be performed in the spatial domain, leading to a substantial improvement in the computational efficiency of graph convolutional models.

By leveraging the power of convolutional filtering techniques, graph convolutional networks (GCNs) have emerged as highly effective models for handling graph-structured data. GCNs have demonstrated remarkable performance in a wide range of tasks involving graphs. One of the key advantages of GCNs is their ability to capture the relational information between nodes in a graph. The convolutional filtering operation allows GCNs to aggregate information from neighboring nodes, enabling them to effectively reason about the global structure and relationships within the graph.

GCNs have been successfully applied in various domains, including social network analysis, recommendation systems, bioinformatics, and knowledge graph reasoning. In social network analysis, GCNs can be used to predict user preferences, identify influential users, or detect communities within a network. In recommendation systems, GCNs have proven effective in generating personalized recommendations by leveraging the user-item interactions represented as graphs. In bioinformatics, GCNs have been applied to tasks such as protein–protein interaction prediction and drug discovery. Additionally, GCNs have been instrumental in knowledge graph reasoning, enabling the inference of missing information and the discovery of latent patterns within complex knowledge graphs [9].

The computational efficiency gained through the simplification of frequency domain convolution has opened up avenues for scalable and efficient modeling of graph data. This advancement has paved the way for further research and innovation in the field of graph convolutional networks, with the potential to revolutionize various domains reliant on graph analysis. In the subsequent years, the advent of frequency domain convolution opened up new possibilities for the development of spatial Graph Convolutional Networks (GCNs), leading to the emergence of numerous variations. These novel approaches collectively fall under the category of Graph Neural Networks (GNNs) [10]. The introduction of GNN models has significantly enhanced their capacity to accommodate diverse types of graph data [11].

Currently, most GNN-based traffic prediction models are built on the basis of convolutional graph networks in the spatial and frequency domains. They construct the topology or semantic relationships between nodes using a graph and use graph convolution to extract structured spatial features from non-Euclidean space data. For traffic networks, the information between different nodes influences each other in the spatial dimension, and the degree of influence between nodes is dynamic and varies at different time steps. In the time dimension, the dependence between different time steps is also different. To capture the dynamics and importance of different nodes and time steps, graph representation learning and attention mechanisms are introduced to determine the weights of different node relationships and time steps, in order to capture richer spatiotemporal correlations. Most traffic prediction tasks are spatiotemporal prediction problems, with inputs consisting of sequences based on spatiotemporal graphs. Therefore, graph spatiotemporal networks integrate modules for processing time series data in convolutional graph networks, such as LSTM, GRU, CNN, etc., in order to capture temporal correlations. Graph spatiotemporal networks can input features after graph convolution processing into time series data modules to capture spatiotemporal correlations. They can also stack multiple time series modules and graph convolution modules in sequence to construct spatiotemporal blocks that capture spatiotemporal correlations. They can also design two independent components to extract temporal and spatial features and integrate spatiotemporal information through tensor dimensional transformation. Table 1.2 compares the four types of graph neural networks.

In practice, CNN models are commonly used for tasks involving spatial data, such as image classification, while RNN models are often preferred for time series problems due to their ability to capture long-term dependencies. However, researchers have conducted extensive studies on utilizing CNN models for learning in spatiotemporal data. To address the challenge of capturing long-term dependencies in time sequences using CNNs, attention-based spatiotemporal graph convolutional networks have been proposed. This network architecture consists of three

Table 1.2 Comparison of graph neural network

Method	Type	Advantage	Disadvantages
Convolutional graph network	Based on spatial domain	Strong flexibility and scalability	Lacks theoretical support
	Based on frequency domain	Solid theoretical foundation	Fixed structure, not suitable for directed graphs
Graph attention networks	/	Adaptively attending to important neighbors	Only focusing on the local structure of nodes
Graph autoencoder	/	Reshaping graphs while preserving dynamic stochasticity involves maintaining the random nature of the data and updates	To learn an effective data distribution form, it is necessary to modify the loss function
Graph spatiotemporal networks	Based on RNN	Widely applicable and with excellent performance	Iteration time-consuming and complex gating mechanisms.
	Based on CNN	Parallel computing and stable gradients	Lack of memory for long sequences

components designed to extract spatiotemporal features by capturing different patterns along the time axis.

These patterns include weekly patterns, daily patterns, and recent intervals of three different types. The network integrates a spatiotemporal attention mechanism with spatiotemporal convolutions, which involve graph convolutions in the spatial dimension and one-dimensional convolutions in the temporal dimension. This combination enables the model to effectively capture dynamic spatiotemporal features of traffic networks. Consequently, it leads to improved prediction accuracy in traffic-related tasks.

By leveraging the power of both attention mechanisms and convolutional operations, this network architecture provides a promising approach for modeling spatiotemporal data and addressing the challenges posed by time series problems.

1.5 The Organizational Structure of This Book

With the rapid development of information technology, the traditional automotive industry is undergoing a profound transformation. Connected vehicles, also known as intelligent connected vehicles, refer to automobiles that achieve comprehensive connectivity between vehicles, people, and infrastructure through advanced sensing, data communication, and control technologies. They provide the traditional functions of transportation and enable real-time exchange and processing of information, greatly enhancing driving safety, efficiency, and comfort. Connected vehicles can collect and analyze traffic data in real-time, optimize route selection, reduce congestion, improve energy efficiency, and help decrease the occurrence of traffic accidents. Moreover, connected vehicles can interact with urban infrastructure, such as intelligent traffic signals, thus realizing more intelligent urban traffic management. Based on the characteristics of connected vehicles, they can share road condition information with other vehicles through the vehicle network to collaboratively avoid collisions. The interaction between vehicles and roadside facilities also offers new urban planning and management possibilities. Through the real-time data collection and analysis of connected vehicles, traffic flow can be predicted more accurately, providing a scientific basis for traffic management and control. Traffic flow prediction based on connected vehicles is a new research direction that combines cutting-edge technologies such as the vehicle network, big data analysis, and artificial intelligence, attracting increasing attention.

Connected vehicles collect real-time traffic data through on-board sensors, communication devices, and other advanced technologies. These data include vehicle location, speed, direction, and other information that can be used for traffic flow prediction. The spatial-temporal variability of real-world net-connected traffic, as well as various factors such as location, weather, and news, has always made traffic flow prediction challenging. First, both historical traffic conditions and adjacent traffic conditions affect the traffic conditions in the prediction area, resulting in traffic flows with different trends in spatial and temporal features. Then, location (downtown and suburban areas), weather (sunny and stormy), news (sporting events and holiday schedules), and other factors can have different impacts on traffic flow. It is crucial to consider the correlation of various factors in the prediction model. Therefore, extracting the spatial-temporal features of the combined external factors is an important method to improve the capability of the connected traffic prediction model.

The book consists of five chapters, beginning with an introductory section that outlines the historical development of artificial intelligence and graph neural networks in the context of connected vehicles. This section also reviews related work on traffic prediction and provides essential background knowledge. Chapters 2 through 5 each focus on a specific prediction method for connected traffic, offering a detailed examination that includes problem definition, model architecture, experimental setup, and a discussion of results, along with relevant references. The concluding chapter summarizes the key contributions of the book and addresses future challenges in the field.

1.6 Conclusion

Graph neural networks have brought transformative advancements to traffic prediction within the realm of connected vehicles (CV), utilizing the adaptable structure of graph data with diverse attribute types to uncover complex spatial-temporal correlations in CV networks. These capabilities have significantly improved the accuracy and effectiveness of traffic prediction models, which are essential for connected vehicle systems. In CV scenarios, addressing both temporal and spatial dimensions is critical for achieving precise and responsive predictions. This can be accomplished by embedding spatial representations into temporal modules or by hierarchically interleaving spatial and temporal components, which indirectly captures the necessary spatiotemporal correlations. Additionally, constructing local spatiotemporal graphs across adjacent time steps allows for the direct modeling of spatiotemporal dependencies, vital for real-time decision-making in connected vehicles.

Furthermore, various temporal features, including traffic flow data patterns, are relevant for CV applications. Predictive data in CV environments not only correlates with immediate real-time data but also exhibits significant periodic patterns influenced by factors like weekdays, holidays, and weekends. Recognizing and integrating these temporal patterns into prediction models are essential for the development of reliable CV systems. By incorporating spatiotemporal attention mechanisms, the model's ability to capture dynamic traffic patterns is enhanced, allowing for an adaptive understanding of evolving traffic conditions in CV networks. Specifically, spatial attention mechanisms capture the influence weights of neighboring nodes, while temporal attention mechanisms focus on key historical sequence data, enabling the model to prioritize important nodes and critical time steps effectively.

In summary, graph-based traffic prediction technology in connected vehicles can be characterized as follows:

1. In-depth exploration of node correlations: By examining nodes like road segments, regions, trajectory points, and origin-destination pairs (OD pairs), the model contributes to a comprehensive understanding of CV network dynamics, essential for coordinated decision-making.
2. Thorough analysis of time-varying characteristics: Through capturing the periodic, pulsatile, and random nature of traffic attributes, models can effectively reflect the dynamic behaviors in CV systems, accommodating real-world traffic fluctuations.
3. Incorporation of external factors: The model integrates various external influences, enhancing its awareness of contextual information (e.g., weather, events) and improving prediction accuracy in the CV framework.
4. Development of graph spatiotemporal network prediction frameworks: These frameworks leverage spatiotemporal correlations and enhance model expressiveness using advanced deep learning techniques like spatiotemporal attention mechanisms, graph representation learning, and multi-task learning, resulting in more accurate, robust, and responsive traffic predictions essential for connected vehicles.

Through these advancements, graph neural networks enable connected vehicle systems to operate with improved efficiency, responsiveness, and safety, ultimately supporting the vision of a seamlessly integrated and intelligent transportation ecosystem.

References

1. K.-H.N. Bui, J. Cho, H. Yi, Spatial-temporal graph neural network for traffic forecasting: An overview and open research issues. Appl. Intell. **52**(3), 2763–2774 (2022). https://doi.org/10.1007/s10489-021-02587-w
2. W.S. McCulloch, W. Pitts, A logical calculus of the ideas immanent in nervous activity. Bull. Math. Biol. **5**(4), 115–133 (1943). https://doi.org/10.1007/bf02478259
3. V. Sze, Y.-H. Chen, T.-J. Yang, J.S. Emer, Efficient processing of deep neural networks: A tutorial and survey. Proc. IEEE Inst. Electr. Electron. Eng. **105**(12), 2295–2329 (2017). https://doi.org/10.1109/jproc.2017.2761740
4. V. Upadhya, P.S. Sastry, An overview of restricted Boltzmann machines. J. Indian Inst. Sci. **99**(2), 225–236 (2019). https://doi.org/10.1007/s41745-019-0102-z
5. O. Abdel-Hamid, A.-R. Mohamed, H. Jiang, G. Penn, Applying convolutional neural networks concepts to hybrid NN-HMM model for speech recognition, in *2012 IEEE International Conference on Acoustics, Speech and Signal Processing (ICASSP)*, (IEEE, 2012)
6. L. Tang, Y. Zhao, J. Cabrera, J. Ma, K.L. Tsui, Forecasting short-term passenger flow: An empirical study on Shenzhen metro. IEEE Trans. Intell. Transp. Syst. **20**(10), 3613–3622 (2019). https://doi.org/10.1109/tits.2018.2879497
7. M.M. Hamed, H.R. Al-Masaeid, Z.M.B. Said, Short-term prediction of traffic volume in urban arterials. J. Transp. Eng. **121**(3), 249–254 (1995). https://doi.org/10.1061/(asce)0733-947x(1995)121:3(249)
8. I. Okutani, Y.J. Stephanedes, Dynamic prediction of traffic volume through Kalman filtering theory. Trans. Res. Part B: Methodol. **18**(1), 1–11 (1984). https://doi.org/10.1016/0191-2615(84)90002-x
9. J. McFadden, W.-T. Yang, S. Durrans, Application of artificial neural networks to predict speeds on two-lane rural highways. Transp. Res. Rec. **1751**(1), 9–17 (2001). https://doi.org/10.3141/1751-02
10. X. Ling, X. Feng, Z. Chen, Y. Xu, H. Zheng, Short-term traffic flow prediction with optimized multi-kernel support vector machine, in *2017 IEEE Congress on Evolutionary Computation (CEC)*, (IEEE, 2017)
11. Y. An, M. Xu, C. Shen, Classification method of teaching resources based on improved KNN algorithm. Int. J. Emerg. Technol. Learn. **14**(04), 73 (2019). https://doi.org/10.3991/ijet.v14i04.10131
12. L. Vanajakshi, L.R. Rilett, A comparison of the performance of artificial. neural networks and support vector machines for the prediction of traffic speed, in *IEEE Intelligent Vehicles Symposium, 2004*, (IEEE, 2004)
13. H. Tan, X. Xuan, Y. Wu, Z. Zhong, B. Ran, A comparison of traffic flow prediction methods based on DBN, in *CICTP 2016*, (American Society of Civil Engineers, Reston, VA, 2016)
14. T. Nguyen, G. Nguyen, B.M. Nguyen, EO-CNN: An enhanced CNN model trained by equilibrium optimization for traffic transportation prediction. Procedia. Comput. Sci. **176**, 800–809 (2020). https://doi.org/10.1016/j.procs.2020.09.075
15. X. Ma, Z. Tao, Y. Wang, H. Yu, Y. Wang, Long short-term memory neural network for traffic speed prediction using remote microwave sensor data. Transp. Res. Part C Emerg. Technol. **54**, 187–197 (2015). https://doi.org/10.1016/j.trc.2015.03.014
16. H. Shao, B.-H. Soong, Traffic flow prediction with long short-term memory networks (LSTMs), in *2016 IEEE Region 10 Conference (TENCON)*, (IEEE, 2016)
17. S. Du, T. Li, X. Gong, Y. Yang, S.J. Horng, Traffic flow forecasting based on hybrid deep learning framework, in *2017 12th International Conference on Intelligent Systems and Knowledge Engineering (ISKE)*, (IEEE, 2017)
18. J. Zhang, Y. Zheng, D. Qi, Deep spatio-temporal residual networks for citywide crowd flows prediction. Proc. Conf. AAAI Artif. Intell. **31**(1) (2017). https://doi.org/10.1609/aaai.v31i1.10735

19. D. Yang, K. Chen, M. Yang, X. Zhao, Urban rail transit passenger flow forecast based on LSTM with enhanced long-term features. IET Intell. Transp. Syst. **13**(10), 1475–1482 (2019). https://doi.org/10.1049/iet-its.2018.5511
20. M. Cao, V.O.K. Li, V.W.S. Chan, A CNN-LSTM model for traffic speed prediction, in *2020 IEEE 91st Vehicular Technology Conference (VTC2020-Spring)*, (IEEE, 2020)
21. Z. Zhene et al., Deep convolutional mesh RNN for urban traffic passenger flows prediction, in *2018 IEEE SmartWorld, Ubiquitous Intelligence & Computing, Advanced & Trusted Computing, Scalable Computing & Communications, Cloud & Big Data Computing, Internet of People and Smart City Innovation (SmartWorld/SCALCOM/UIC/ATC/CBDCom/IOP/SCI)*, (IEEE, 2018)
22. Y. Yu, J. Chen, T. Gao, M. Yu, DAG-GNN: DAG structure learning with graph neural networks. arXiv [cs.LG] (2019) http://arxiv.org/abs/1904.10098
23. T. Zhang, W. Ding, T. Chen, Z. Wang, J. Chen, A graph convolutional method for traffic flow prediction in highway network. Wirel. Commun. Mob. Comput. **2021**, 1–8 (2021). https://doi.org/10.1155/2021/1997212
24. Y. Han, T. Peng, C. Wang, Z. Zhang, G. Chen, A hybrid GLM model for predicting citywide spatio-temporal metro passenger flow. ISPRS Int. J. Geoinf. **10**(4), 222 (2021). https://doi.org/10.3390/ijgi10040222
25. S. Feng, J. Ke, H. Yang, J. Ye, A multi-task matrix factorized graph neural network for co-prediction of zone-based and OD-based ride-hailing demand. IEEE Trans. Intell. Transp. Syst. **23**(6), 5704–5716 (2022). https://doi.org/10.1109/tits.2021.3056415
26. C. Song, Y. Lin, S. Guo, H. Wan, Spatial-temporal synchronous graph convolutional networks: A new framework for spatial-temporal network data forecasting. Proc. Conf. AAAI Artif. Intell. **34**(01), 914–921 (2020). https://doi.org/10.1609/aaai.v34i01.5438

Chapter 2
Spatial-Temporal Graph Convolutional Networks for Connected Traffic Prediction

2.1 Complex Correlation Features of Road Nodes

Traditional method uses the adjacency matrix to describe the relationship between nodes, ignoring the influence of distance, correlation, and external interference between nodes. As shown in Fig. 2.1a, the linear distance between node A and node B is the closest, but there is no direct connection, so the correlation between the two nodes does not increase with the decrease in distance. As shown in Fig. 2.1b, there are similar traffic flow changes between node A and node B, so it is essential to analyze the data correlation between the two nodes. As shown in Fig. 2.1c, node A has different traffic flow changes in different weather, so it is crucial to analyze the external impact of the traffic environment. Therefore, only using a single adjacency matrix to reflect the relationship between nodes is not conducive to improving the performance of the traffic flow prediction model.

Q. Shi et al., *Connected Vehicles Traffic Prediction*, Wireless Networks,
https://doi.org/10.1007/978-3-031-84548-2_2

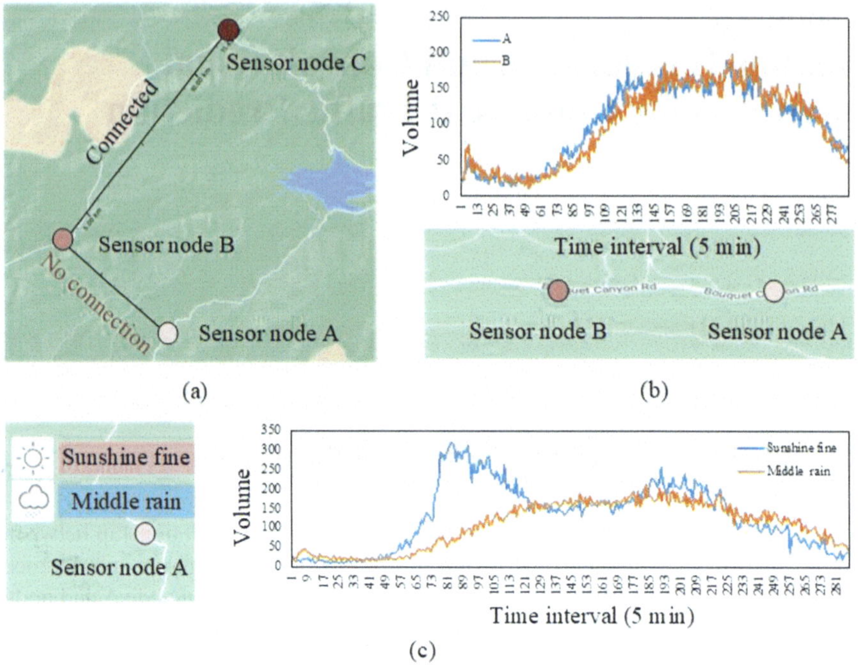

Fig. 2.1 Complex correlation features of road nodes. (**a**) Influence of distance. (**b**) Influence of data correlation. (**c**) Influence of external factors

2.2 Static and Dynamic Spatial Correlation Neural Network (SDSCNN)

Many works use ChebNet or GCN to analyze the spatial correlation. However, whether it is ChebNet or GCN, there are three shortcomings:

1. From the history of GCN-related research, GCN and its original model ChebNet are both models of the spectral domain, which uses graph theory to perform convolution operations on topological graphs. Convolution operations on topological graphs (graph Fourier transform and graph convolution) require the use of Laplacian matrices. The Laplace matrix of an undirected graph is a positive semi-definite matrix. The spectral decomposition of the positive semi-definite matrix can be decomposed into N linearly independent eigenvectors, to carry out the subsequent graph Fourier transformation and convolution, and finally obtain the first-order derivation formula of GCN. However, the Laplacian matrix of a directed graph is not a positive semi-definite matrix, so spectral decomposition is not possible. Theoretically, the derivation formula of GCN cannot be obtained.

2. GCN cannot handle dynamic graphs. GCN relies on the specific graph structure during training, and it is also performed on the same graph during testing. It is a full graph calculation method. GCN can only handle transudative learning tasks and cannot be applied to inductive tasks.
3. GCN cannot assign different weights to each neighbor. GCN treats all neighbor nodes uniformly during convolution and cannot assign different weights according to the importance of nodes.

Based on the above issues, GAT is proposed. There are at least three benefits of using GAT rather than using GCNs.

1. GAT is a spatial domain model that does not require the Laplacian matrix to join operations. Therefore, it is more suitable for directed graphs.
2. The operation mode of GAT is node-wise operation, instead of GCN using the Laplacian matrix for convolution. Therefore, it can be used for dynamic graphs.
3. The spatial weights of GAT are computed dynamically, instead of GCN utilizing the Laplacian matrix. Therefore, GAT allows for assigning different weights to nodes of the same neighborhood.

We propose a hybrid model integrating static and dynamic spatial correlation neural network for traffic prediction. Figure 2.2 shows the model structure proposed in this paper. The model divides the spatial correlation into two parts, including the static spatial correlation analysis and the dynamic spatial correlation analysis. Specifically, the static spatial correlation is constructed according to the static adjacency matrix and GAT. Then the output of the above two parts is summed as the final output of the spatial correlation analysis. For the temporal correlation part, we use the multi-head self-attention to get the output. The above network structure is constructed as an SDSCNN layer, and we stacked multiple SDSCNN layers in our model for further exploration. Finally, the output of the last SDSCNN layer is fed into the multi-layer perceptron (MLP) layer to predict the specific moment traffic prediction value. The model includes four parts: static spatial module, dynamic spatial module, spatial fusion module, and temporal module. The specific description is as below.

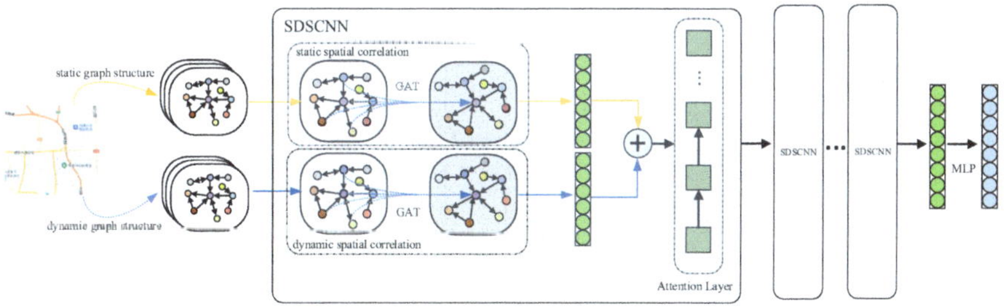

Fig. 2.2 Structure of SDSCNN

2.2.1 Static Spatial Module

GAT introduces a self-attention mechanism in the propagation process, and the hidden state of each node is calculated by paying attention to its neighbor nodes. The GAT network is implemented by stacking a simple graph attention layer. For each attention layer to nodes, the attention coefficient calculation method is as shown in Eq. (2.1):

$$\alpha_{ij} = \frac{\exp\left(\mathrm{LeakyReLU}\left(\boldsymbol{a}^T\left[\boldsymbol{W}\boldsymbol{h}_i \parallel \boldsymbol{W}\boldsymbol{h}_j\right]\right)\right)}{\sum_{k \in N_i} \exp\left(\mathrm{LeakyReLU}\left(\boldsymbol{a}^T\left[\boldsymbol{W}\boldsymbol{h}_i \parallel \boldsymbol{W}\boldsymbol{h}_k\right]\right)\right)} \tag{2.1}$$

Among them, α_{ij} is the attention coefficient from node i to j, and N_i represents the neighbor node of node i. The input feature of the node is $\boldsymbol{h} = \{h_1, h_2, \ldots, h_N\}$, $h_i \in R^F$, where N, F represents the number of nodes and the feature dimension, respectively. $\boldsymbol{h}$ is the feature value of the nodes. $\boldsymbol{W} \in R^{F'}$ is the linear transformation weight matrix applied on each node, and $a \in R^{2F'}$ is the weight vector, which can map the input to R. Finally, softmax is used for normalization, and LeakyReLU is added for activation.

The feature output of the final node is obtained by Eq. (2.2):

$$\boldsymbol{h}_i' = \sigma\left(\sum_{j \in N_i} \alpha_{ij} \boldsymbol{W}\boldsymbol{h}_j\right) \tag{2.2}$$

Among them, α_{ij} is the attention coefficient of node i to j calculated in Eq. (2.3), $\boldsymbol{W}$ is the weight matrix to be trained, $\boldsymbol{h}_j$ is the feature of the neighbor node j of node i, and $\boldsymbol{h}_i'$ is the node i calculated by the attention mechanism output.

In addition, GAT utilizes multi-head self-attention to stabilize the learning process. It applies an independent attention mechanism to calculate the hidden state. For the results of the self-attention layers, this method concatenates the output and feeds into the fully connected layer to get the final result. The process is shown in Fig. 2.3, Eq. (2.3) is the specific calculation process of multi-head attention:

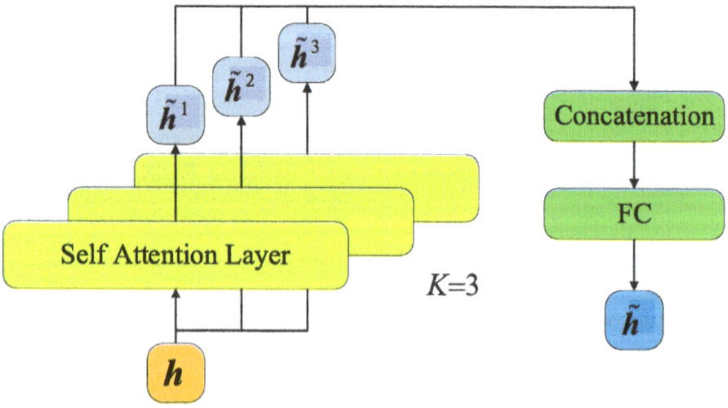

Fig. 2.3 Illustration of multi-head self-attention

$$h_i' = \|_{k=1}^{K} \sigma\left(\sum_{j \in N_i} \alpha_{ij}^k \boldsymbol{W}^k \boldsymbol{h}_j\right) \tag{2.3}$$

where α_{ij}^k is the normalized attention coefficient of the first attention head and || represents the concatenation operation.

Our model introduces the multi-head self-attention method to mine the correlation. Static spatial correlation analysis depends on the original road topology. Generally existing work uses a 0–1 matrix to represent the road structure. However, it cannot accurately reflect the static spatial structure of the road network. Therefore, this paper uses the adjacency matrix based on the distance from the road network, which not only better represents the actual spatial structure of the road network, but also reduces the computational complexity of the model. And the product of adjacency matrix and parameter matrix is used as the input of GAT. The static adjacency matrix calculation formulas are as follows:

$$\tilde{\boldsymbol{A}}_{ij} = \begin{cases} e^{-\frac{\mathrm{dist}(v_i, v_j)^2}{\sigma^2}}, & \mathrm{dist}(v_i, v_j) < d^g \\ 0, & \mathrm{dist}(v_i, v_j) \geqslant d^g \end{cases} \tag{2.4}$$

where the adjacency matrix $\tilde{\boldsymbol{A}} \in \mathbb{R}^{N \times N}$ is calculated using a threshold Gaussian kernel and the road distance between each node of the road network. $\mathrm{dist}(v_i, v_j)$ represents the distance between sensor v_i and sensor v_j, σ^2 is the variance of the road distance, and d^g is the threshold. Since the actual road network is generally a directed graph, $\mathrm{dist}(v_i, v_j)$ and $\mathrm{dist}(v_j, v_i)$ are inconsistent in some cases, so $\boldsymbol{A}$ is an asymmetric matrix. d^g is the maximum threshold of road network distance, and different datasets have different values.

The feature value X and the matrix $\boldsymbol{W}^s$ are input into Eq. (2.4) to obtain the static spatial correlation analysis result. The calculation rule is as follows:

$$\boldsymbol{X}^s = \|_{k=1}^{K} \sigma\left(\sum_{j \in N_i} \alpha_{ij}^{k^s} \boldsymbol{W}^s \boldsymbol{X}\right) \tag{2.5}$$

$$\boldsymbol{W}^s = \tilde{\boldsymbol{A}} \boldsymbol{\theta}^s \tag{2.6}$$

where $\boldsymbol{X}^s$ is the output of static spatial correlation analysis. $\boldsymbol{\theta}^s \in \mathrm{R}^{N \times N}$ is a learnable parameter. Sometimes, the adjacency matrix $\boldsymbol{A}$ of the road distance cannot be directly used in the traffic flow prediction task. We consider the impact of road construction, natural disasters, etc., and use the parameter $\boldsymbol{\theta}^s$ to get the matrix $\boldsymbol{W}^s$. $\alpha_{ij}^{k^s}$ is calculated by Eq. (2.1), and keeps fixed during the calculation. It represents the weights of the edges trained by the GAT. The product $\alpha_{ij}^{k^s}$ and $\boldsymbol{W}^s$ determines the relationship of each node in the static spatial correlation module. The edges and the associated weights of the nodes in this model are static.

2.2.2 Dynamic Spatial Module

Dynamic spatial correlation often exists in the historical data. Similarly, we also use the multi-headed attention mechanism for dynamic spatial correlation mining. The difference is that the attention coefficients of each layer changes according to the input feature values.

Following the same idea as the static weight calculation, the dynamic part uses the fusion of two weight matrices to construct the dynamic weights. One is the matrix $\boldsymbol{W}^d$, and the other is the matrix $\alpha_{ij}^{k^d}$ which is calculated by GAT.

The weight parameters depend on the current feature values, and the correlation between nodes at different moments shows dynamic characteristics, so we introduce the dynamic weight training method to construct the $\boldsymbol{W}^d$. The optimal dynamic feature at the current moment is determined by combining the latest feature of the nodes and the weight matrix of the model which is self-trained in real time. So, we first parameterize the matrix $\boldsymbol{W}_\phi$. And then the matrix $\boldsymbol{W}_\phi$ and the latest feature value $\boldsymbol{X}$ are input into the heuristic function, the specific formula is as follows:

$$\boldsymbol{W}^d = \phi\left(\boldsymbol{X}, \boldsymbol{W}_\phi\right) = \boldsymbol{X}^T \boldsymbol{W}_\phi \boldsymbol{X} \tag{2.7}$$

where $\phi(\cdot)$ is the heuristic function. $\boldsymbol{W}^d$ is the output of dynamic spatial correlation analysis.

For the $\alpha_{ij}^{k^d}$, the latest values of the nodes are input into Eq. (2.15) to get the output.

Also, the dynamic spatial correlation is analyzed using the GAT which is calculated as follows:

$$\boldsymbol{X}^d =_{k=1}^{K} \sigma\left(\sum_{j \in N_i} \alpha_{ij}^{k^d} \boldsymbol{W}^d \boldsymbol{X}\right) \tag{2.8}$$

where $\alpha_{ij}^{k^d}$ is calculated by Eq. (2.1), and the input is the latest features of the nodes. It represents the training parameters of the dynamic spatial module. $\boldsymbol{X}^d$ is the output of dynamic spatial correlation analysis.

2.2.3 Spatial Fusion Module

Spatial correlation analysis includes static spatial correlation analysis and dynamic spatial correlation analysis. We add both of them to achieve the fusion of spatial correlation. Therefore, based on the above analysis, the spatial correlation result is as follows:

$$\tilde{\boldsymbol{X}} = \boldsymbol{X}^s + \boldsymbol{X}^d \tag{2.9}$$

where $\tilde{\boldsymbol{X}}$ is the result of the spatial correlation.

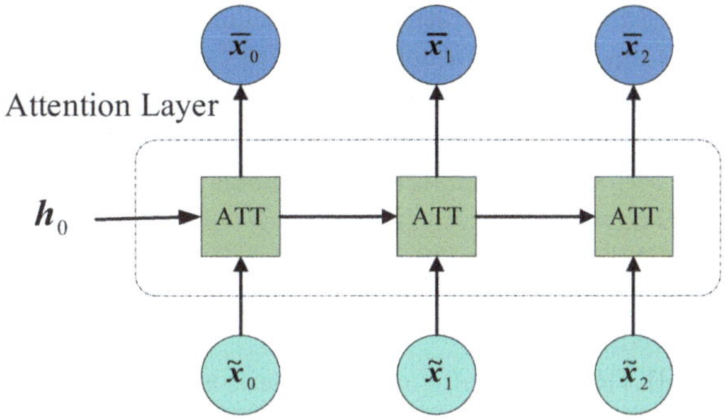

Fig. 2.4 Illustration of the attention layer

2.2.4 *Temporal Module*

To improve the model's ability of mining temporal correlation, especially the periodicity, the multi-head self-attention mechanism is also used in this module. Since the self-attention mechanism is to calculate the attention at every time point and all time points, it is possible to directly calculate the dependency relationship regardless of the interval between time points. The model structure is shown in Fig. 2.4 and the input of the model is X . The calculation rule is as below:

$$\bar{X} =_{k=1}^{K} \sigma\left(\sum_{j\in N_i} \alpha_{ij}^{k} W^{k} \tilde{X}\right) \tag{2.10}$$

where α_{ij}^{k} is calculated by Eq. (2.1) and the input is the current node's value. $\bar{X}$ is the result of the temporal correlation. Based on multi-head self-attention, this paper constructs a temporal correlation analysis module of Seq2Seq, which can be extracted in different periods to predict information of different time interval lengths.

2.2.5 *Experiments*

To evaluate the effectiveness of the proposed model, a series of experiments have been conducted. They are organized into the following steps.

2.2.5.1 Experiments Settings

Data Collection

The SDSCNN in this paper is validated using three traffic datasets. The highway datasets PEMS04, PEMS08, and METR-LA in the Caltrans Performance Evaluation System (PeMS, https://pems.dot.ca.gov) are used for analysis.

PEMS04 is the traffic data collected from San Francisco Bay, which contains 3848 sensors on 29 roads. The PEMS dataset contains traffic information such as traffic flow, vehicle speed, and density. In addition to this, the road distance between the sensors is also included. Among them, 307 sensors are selected for prediction. The period of this dataset is from January 1, 2018, to February 28, 2018, with a total of 59 days of data.

PEMS08 is the traffic data of San Bernardino for 62 days from July 1, 2016, to August 31, 2016. It contains 1979 sensors on 8 roads, and 170 of them are selected for prediction.

METR-LA is traffic data for the Los Angeles Freeway from March 1, 2012, to June 27, 2012, with 207 sensors. METR-LA dataset mainly records vehicle speed information but also includes the GPS location of the sensors and the road distance between the sensors.

Evaluation Metric

This article uses three measurement functions to evaluate the prediction effect of the model, namely Mean Absolute Error (MAE), Root Mean Square Error (RMSE), and Mean Absolute Percentage Error (MAPE).

$$\mathrm{MAE}=\frac{1}{MN}\sum_{i=1}^{M}\sum_{j=1}^{N}\left|y_j^i-\hat{y}_j^i\right| \tag{2.11}$$

$$\mathrm{RMSE}=\sqrt{\frac{1}{MN}\sum_{i=1}^{M}\sum_{j=1}^{N}\left(y_j^i-\hat{y}_j^i\right)^2} \tag{2.12}$$

$$\mathrm{MAPE}=\frac{1}{M}\sum_{i=1}^{M}\sum_{j=1}^{N}\left|\frac{y_j^i-\hat{y}_j^i}{y_j^i}\right| \tag{2.13}$$

where yji and ŷji represent the true and predicted values, respectively, M represents the number of samples, and N represents the number of roads.

Model Parameter Settings

This paper builds a new SDSCNN model based on the deep learning framework PyTorch and uses two NVIDIA TITAN 12GB GPUs for experiments. In the experiment, the mean absolute error (MAE) was used as the Loss function and used Adam optimizer to optimize the model parameters. The optimizer training step size is set to 0.0001, the number of batches is set to 40, the maximum number of iterations is set to 500, and the early_stop is set to 10. Also, initialize the training parameters with a random function. In this paper, grid search is used to optimize some hyper-parameters. The recommended hyper-parameters are shown in Table 2.1. The *k*-fold

Table 2.1 Recommended hyper-parameter settings for the SDSCNN model

Type	Setting
Time steps (T)	8
Layers	3
Head of multi-head attention	8,6,5

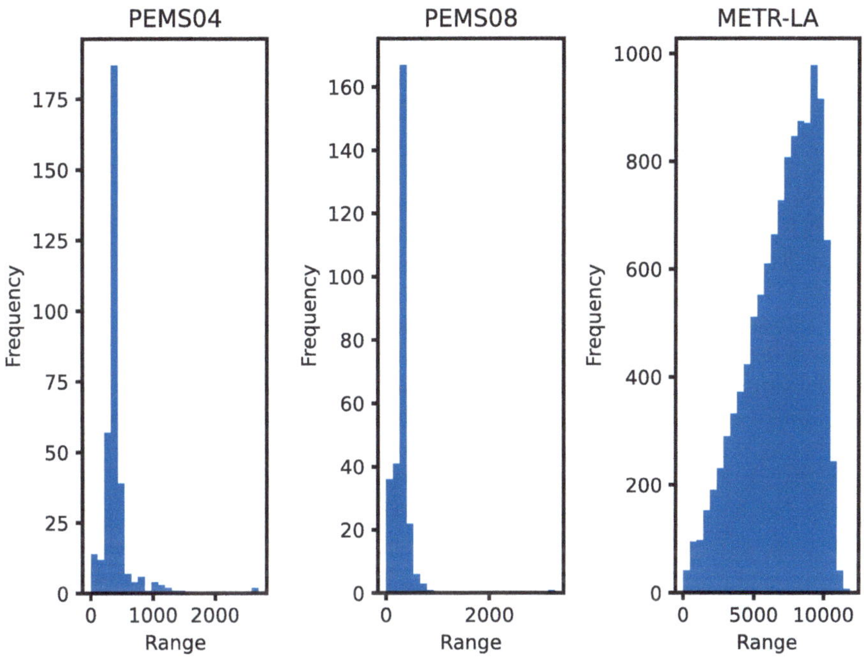

Fig. 2.5 Histogram of PEMS04, PEMS08, and METR-LA road network distances

cross-validation was selected as the dataset division method, and the hyper-parameter cross-validation in Table 2.1 was used to repeat 10 times, and the mean and standard deviation were used as the results.

In addition, the adjacency matrix $A \in \mathbb{R}^{N \times N}$ in Eq. (2.19) uses d^g as the maximum threshold of the road network distance, and different datasets have different values. Figure 2.5 shows the histograms of the road network distances for each dataset. According to the histograms, this paper sets the d^g of PEMS04, PEMS08, and METR-LA to 1500, 1000, and 12,000, respectively.

2.2.5.2 Methods for Comparison

We first compare the proposed SDSCNN method with two model-driven methods, including ARIMA [1] and SVR [2]. We further compare the SDSCNN with five data-driven methods, i.e., (1) LSTM [3], (2) DCRNN [4], (3) Spatial-temporal graph convolutional network (STGCN) [5], (4) Graph WaveNet (GWN) [6], and (5) SLCNN [7].

ARIMA: It is a classic time series data prediction method. In the experiments, we first check the data stability and deal with non-stationary data. We input the data into the ARIMA model and check the model parameters. After the model parameters are checked for reasonableness, the prediction is performed and the results will be output.

SVR: This method is an application of the support vector machine (SVM) in regression tasks. In the experiments, we choose the linear function as the kernel function. And the regularization coefficient is set to 1.

LSTM: This method is a variant of the recurrent neural network. It only considers the temporal correlation. The LSTM uses the gating mechanism, including input gates, output gates, and forgetting gates. We input the traffic flow data into the LSTM to obtain the prediction results.

DCRNN: The DCRNN model represents the road network as a weighted directed graph. The model first performs a diffusion convolution operation of the graph signal and the filter to map PD features to QD features which mines the spatial correlation of traffic flow. Meanwhile, the output is used as the input of GRU to further analyze the temporal correlation. To perform better multi-step prediction, a Seq2Seq model structure is built by diffusion convolution and GRU network. We predict by inputting the traffic flow data into the encoder and initializing the decoder with its final state.

STGCN: STGCN consists of several spatial-temporal convolution modules. Each module contains two gated sequence convolution layers and one spatial graph convolution module. Gated sequence convolution is used to capture temporal correlation and consists of a 1D convolution and a GLU. Compared with the RNN series models, this method reduces the computational complexity and simplifies the model training. The spatial correlation is analyzed by Chebyshev graph convolution.

GWN: The GWN model stacks multiple spatial-temporal layers. A spatial-temporal layer consists of a GCN and a gated TCN. Each layer contains residual connections. The adaptive adjacency matrix is employed in GCN. Meanwhile, the diffusion convolution is extended to graph convolution to represent diffusion graph convolution, which complements the spatial structure of road network to a certain degree. GWN uses causal convolution and dilated convolution to convert all inputs into one-dimensional vectors according to the time dimension and performs 1D convolution operations.

SLCNN: SLCNN is mainly used to address the problem of analyzing dynamic spatial correlation. SLCNN consists of multiple SLCNN layers. Each SLCNN layer includes a Global SLC module and a local SLC module. The global SLC module combines Chebyshev graph convolution with static adjacency matrix and dynamic adjacency matrix, respectively, to achieve global spatial correlation analysis. The local SLC module only considers the spatial correlation between the k-nearest neighbor nodes. Both modules use P3D-SLC networks for the analysis of temporal correlation.

2.2.5.3 Results of Traffic Prediction

The experimental results are shown in Table 2.2. In the PEMS08 dataset, the SDSCNN model is always better than other benchmark models in accuracy. In the PEMS04 dataset, SDSCNN has the smallest MAE and MAPE, and the RMSE is slightly larger than SLCNN. Taking MAPE as the evaluation index, SDSCNN is 6.52% higher than DCRNN, 1.78% higher than STGCN, 17.11% higher than GWN, and 4.02% higher than SLCNN.

Meanwhile, we visualize and compare the prediction results of the SLCNN model with that of the SDSCNN model, as shown in Fig. 2.6. The fit of the SDSCNN model is better than SLCNN. Although there is a certain regularity in the prediction results of SLCNN, there is still a certain distance from the accuracy of the SDSCNN model. This proves that the SDSCNN model adopts the multi-head attention mechanism to replace the 1D convolution of SLCNN and obtains a certain performance improvement. In addition, SDSCNN shows stronger adaptability when there are sudden and large changes in the traffic flow state. This shows that when compared with the global and local graph structures of SLCNN, SDSCNN can better capture the dynamic transformation of the traffic flow by combining static and dynamic graph structures.

2.2.5.4 Results Discussion

Spatial Correlation Analysis

SDSCNN converts the distances between sensors into static spatial correlations, as shown in Fig. 2.7a. From the perspective of road network structure, the structure can distinguish adjacent road segments and non-adjacent road segments, to provide a reference for prediction. Figure 2.7b–d represent the dynamic spatial correlation coefficients trained from the first layer of the SCSCNN layer to the last layer.

We define HCN as the number of correlation coefficients greater than 0.5 in the correlation matrix. From Fig. 2.7, we can be found that the correlation in Fig. 2.7a is not similar to that in Fig. 2.7b–d, which indicates that the real road network structure cannot fully express the spatial correlation, we can calculate the previous traffic flow further improves the model's ability to capture spatial correlations.

The density of strong correlation coefficients in Fig. 2.7b is lower than that in Fig. 2.7a, c, d. This is because the first hidden layer of SDSCNN can only obtain information about adjacent nodes for analysis, so the density of strong correlation coefficients is not as large as shown in Fig. 2.7a. When the information aggregated by the previously hidden layer of SDSCNN is used as the input of the next hidden layer, the next hidden layer can obtain more information about non-adjacent nodes. Therefore, the density of strong correlation coefficients in Fig. 2.7c, d is higher than that in Fig. 2.7b.

Table 2.2 Performance comparison of traffic flow prediction of different models

Methods									
Datasets	Metrics	ARIMA	SVR	LSTM	DCRNN	STGCN	GWN	SLCNN	SDSCNN
PEMS04	MAE	32.11 ± 0.09	28.27 ± 0.01	24.38 ± 0.35	24.48 ± 0.30	22.7 ± 0.64	25.45 ± 0.03	22.56 ± 0.51	**21.13 ± 0.35**
	RMSE	68.13 ± 0.16	44.56 ± 0.01	40.51 ± 0.37	38.43 ± 0.19	35.55 ± 0.75	39.7 ± 0.04	33.58 ± 0.76	**33.19 ± 0.46**
	MAPE(%)	17.96 ± 0.03	19.20 ± 0.01	16.29 ± 0.38	15.33 ± 0.33	14.59 ± 0.21	17.29 ± 0.24	14.93 ± 0.62	**14.33 ± 0.32**
PEMS08	MAE	24.04 ± 0.08	23.25 ± 0.01	19.07 ± 0.31	20.07 ± 0.27	18.02 ± 0.14	19.13 ± 0.08	17.13 ± 0.76	**16.95 ± 0.34**
	RMSE	43.3 ± 0.12	36.16 ± 0.11	35.83 ± 0.39	31.33 ± 0.22	27.83 ± 0.20	31.05 ± 0.07	**26.8 ± 0.85**	28.35 ± 0.39
	MAPE(%)	15.42 ± 0.07	14.64 ± 0.02	13.09 ± 0.32	12.52 ± 0.29	11.4 ± 0.10	12.68 ± 0.57	10.96 ± 0.80	**10.95 ± 0.33**
METR-LA	MAE	4.01 ± 0.03	3.85 ± 0.05	3.77 ± 0.23	3.15 ± 0.21	2.87 ± 0.06	3.14 ± 0.05	2.88 ± 0.49	**2.81 ± 0.19**
	RMSE	6.96 ± 0.06	6.57 ± 0.11	6.32 ± 0.29	6.45 ± 0.32	5.54 ± 0.08	5.03 ± 0.03	4.87 ± 0.58	**4.66 ± 0.21**
	MAPE(%)	10.18 ± 0.03	9.54 ± 0.06	9.07 ± 0.20	8.80 ± 0.25	7.43 ± 0.05	7.29 ± 0.35	6.76 ± 0.47	**6.29 ± 0.16**

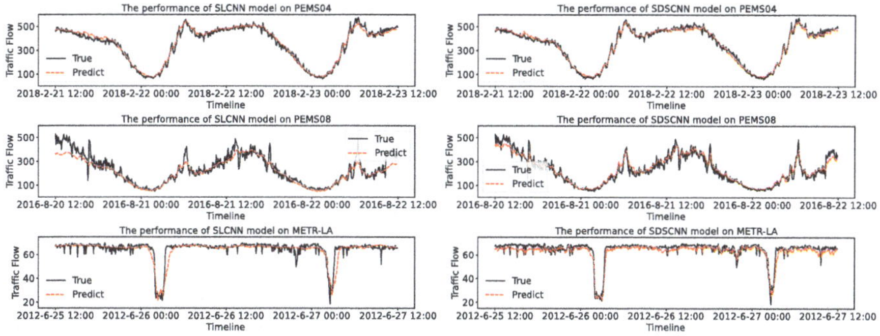

Fig. 2.6 Visualization of prediction results of SLCNN and SDSCNN

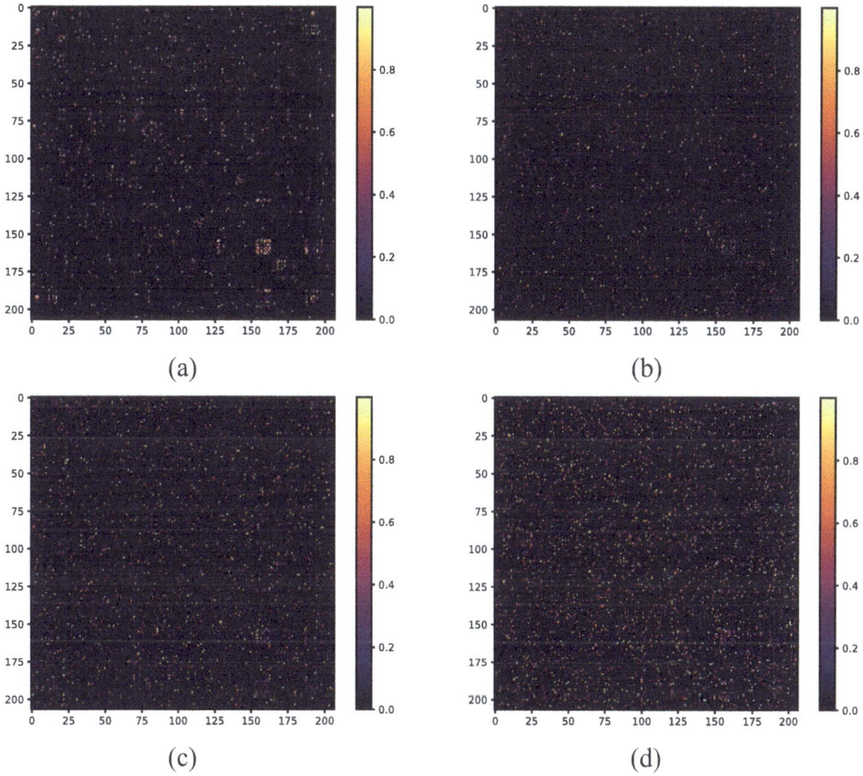

Fig. 2.7 Sensor spatial correlation analysis. (**a**) *HCN*:898. (**b**) *HCN*:591. (**c**) *HCN*:1027. (**d**) *HCN*:1936

Multi-Step Prediction Analysis

Table 2.3 and Fig. 2.8 show the change in the prediction performance of various methods as the prediction output step increases. In general, as the prediction output step gets longer, the corresponding difficulty of the prediction gets greater, so the

Table 2.3 RMSE of different output steps on PEMS04, PEMS08, and METR-LA

Methods									
Datasets	Output steps	ARIMA	SVR	LSTM	DCRNN	STGCN	GWN	SLCNN	SDSCNN
PEMS04	RMSE(5 min)	68.13 ± 0.16	44.56 ± 0.01	40.51 ± 0.37	38.43 ± 0.19	35.55 ± 0.75	39.7 ± 0.04	33.58 ± 0.76	**33.19 ± 0.46**
	RMSE(10 min)	76.65 ± 0.23	52.41 ± 0.04	45.63 ± 0.42	45.63 ± 0.18	38.12 ± 0.93	44.46 ± 0.42	37.65 ± 0.90	**36.45 ± 0.54**
	RMSE(15 min)	95.23 ± 0.27	78.5 ± 0.02	52.32 ± 0.53	52.32 ± 0.30	41.96 ± 1.23	49.83 ± 0.77	41.96 ± 1.18	**39.87 ± 0.61**
PEMS08	RMSE(5 min)	43.3 ± 0.12	36.16 ± 0.11	35.83 ± 0.39	31.33 ± 0.22	27.83 ± 0.20	31.05 ± 0.07	**26.8 ± 0.85**	28.35 ± 0.39
	RMSE(10 min)	49.16 ± 0.21	42.95 ± 0.11	40.53 ± 0.54	40.53 ± 0.36	33.6 ± 0.84	35.63 ± 0.55	33.32 ± 1.04	**31.62 ± 0.51**
	RMSE(15 min)	59.21 ± 0.25	50.54 ± 0.15	46.21 ± 0.70	46.21 ± 0.61	38.51 ± 0.92	40.11 ± 0.83	39.02 ± 1.31	**36.19 ± 0.74**
METR-LA	RMSE(5 min)	6.96 ± 0.06	6.57 ± 0.11	6.32 ± 0.29	6.45 ± 0.32	5.54 ± 0.08	5.03 ± 0.03	4.87 ± 0.58	**4.66 ± 0.21**
	RMSE(10 min)	7.89 ± 0.15	7.12 ± 0.13	6.92 ± 0.37	6.92 ± 0.48	6.58 ± 0.51	5.56 ± 0.28	5.37 ± 0.64	**5.23 ± 0.44**
	RMSE(15 min)	8.51 ± 0.22	8.35 ± 0.15	7.88 ± 0.50	7.88 ± 0.69	7.01 ± 0.78	6.36 ± 0.49	5.97 ± 0.88	**5.81 ± 0.59**

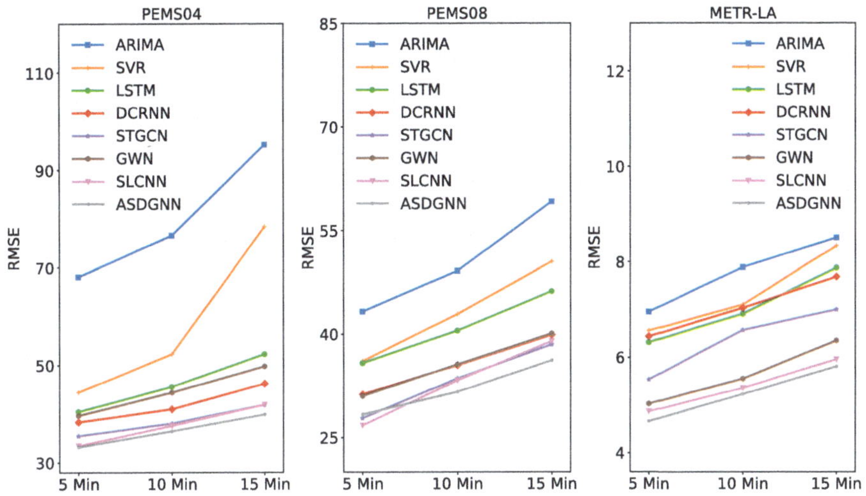

Fig. 2.8 RMSE of different output steps on PEMS04, PEMS08, and METR-LA

prediction error also increases. As can be seen from the images, methods that only consider temporal correlation can achieve good results in short-term prediction, such as ARIMA, SVR, and LSTM. However, their forecasting performance dropped sharply as the forecast output step increased. In contrast, graph-based methods degrade slowly. This is mainly because they take into account the traffic flow effects of adjacent areas and the previous period. The prediction errors of graph-based methods increase slowly with the output step, and they all perform well overall.

SDSCNN model achieves the best prediction performance in almost every scenario, except the PEMS08 dataset. We carefully analyzed the PEMS08 dataset and found that the dataset is noisy with some extreme values. The RMSE is more sensitive to noise data, so the error rises slightly when the prediction output step is 5. Moreover, compared with the METR-LA dataset, the PEMS series data are more volatile and have many sudden flow changes, which makes the prediction task more arduous, and as a result, their RMSE values will be larger.

Especially, in multi-step prediction tasks, the difference between SDSCNN and other baselines is more significant, indicating that the strategy combining graph and attention mechanism can better mine the dynamic spatial-temporal patterns of traffic data.

Hyper-Parameter Optimization

We also compare the performance of SDSCNN with different hyper-parameters, including the influence of the time step, the number of layers of SDSCNN, and the number of heads of the multi-head self-attention mechanism on MAPE. The experimental results are shown in Tables 2.4 and 2.5.

Table 2.4 MAPE of different hyper-parameters on PEMS04, PEMS08, and METR-LA

Time steps						
Datasets	Layers	6	7	8	9	10
PEMS04	1	17.27 ± 2.40	17.00 ± 1.22	16.44 ± 1.00	15.91 ± 0.93	16.02 ± 2.07
	2	17.56 ± 0.11	16.91 ± 0.54	14.77 ± 0.61	15.48 ± 1.49	16.50 ± 0.90
	3	16.77 ± 0.72	16.03 ± 0.68	**14.33 ± 0.32**	15.96 ± 0.28	15.65 ± 2.05
	4	18.11 ± 1.30	16.70 ± 1.45	15.40 ± 0.60	16.01 ± 1.18	16.29 ± 0.67
PEMS08	1	13.28 ± 2.41	12.32 ± 1.30	13.72 ± 0.46	12.45 ± 0.72	13.14 ± 1.52
	2	14.22 ± 1.25	12.28 ± 1.25	11.22 ± 0.41	12.28 ± 0.73	12.53 ± 1.30
	3	13.94 ± 0.69	12.73 ± 1.63	**10.95 ± 0.33**	12.01 ± 1.24	13.06 ± 1.70
	4	13.13 ± 2.02	13.41 ± 1.87	11.01 ± 0.26	12.59 ± 1.37	12.82 ± 1.49
METR-LA	1	9.38 ± 2.00	6.56 ± 0.13	6.83 ± 0.36	7.03 ± 0.62	7.88 ± 0.97
	2	8.66 ± 1.56	**5.27 ± 0.88**	5.46 ± 0.06	7.24 ± 1.40	7.10 ± 0.27
	3	9.37 ± 1.55	5.76 ± 1.90	5.81 ± 0.59	7.16 ± 0.52	7.06 ± 1.20
	4	7.99 ± 2.10	6.01 ± 1.30	6.61 ± 0.75	6.90 ± 0.57	7.47 ± 1.93

From Table 2.4, we can find that with the increase of the time step and the increase of the number of SDSCNN layers, the MAPE decreases first and then increases. For different datasets, the optimal hyper-parameters of SDSCNN are different. As can be seen in Table 2.4, the PEMS04 and PEMS08 datasets perform well when the time step is 8 and the number of SDSCNN layers is 3. However, in the METR-LA dataset, our method achieves the best results when the time step is 7 and the number of SDSCNN layers is 2. This may be caused by the inconsistency between the sensor types of METR-LA and the PEMS04 and PEMS08 datasets. In Table 2.5, as the number of heads of the multi-head self-attention mechanism increases, the MAPE tends to decrease slightly. When the number of heads is greater than 5, the accuracy of the model no longer improves. This result is reasonable because the multi-head self-attention mechanism will increase the parameter and increase the risk of overfitting problems.

Ablation Experiment

To further analyze the performance of our model, we added the other two variants under different combinations of modules. These two variants are compared with the SLCNN model and the SDSCNN model in the PEMS04 dataset, and the four models differ in:

(a) SLCNN (GL + 1D): the basic model which uses global and local modules for spatial convolution and 1D convolution to extract temporal dependencies.
(b) SD + 1D: The spatial correlation analysis method of this model is the same as the method of our model. And the temporal module uses the 1D CNN.
(c) GL + ATT: The temporal module of this variant is designed based on a multi-head attention mechanism rather than 1D CNN.

Table 2.5 MAPE of different heads for the multi-head self-attention mechanism on PEMS04, PEMS08, and METR-LA

Datasets / Heads	2	3	4	5	6	7
PEMS04	14.72 ± 0.34	14.55 ± 0.05	14.46 ± 0.34	14.36 ± 0.35	14.43 ± 0.15	14.41 ± 0.10
PEMS08	11.31 ± 0.16	11.24 ± 0.35	11.09 ± 0.48	10.96 ± 0.13	11.01 ± 0.13	10.96 ± 0.18
METR-LA	5.62 ± 0.10	5.50 ± 0.06	5.47 ± 0.06	5.35 ± 0.26	5.32 ± 0.37	5.33 ± 0.11

(d) SDSCNN (SD + ATT): The SDSCNN model deploys a multi-layer spatial module and a temporal module based on a multi-head attention mechanism for prediction.

As shown in Fig. 2.9, the model with static and dynamic spatial modules (SD) achieves a relatively better prediction performance than the model with global and local modules (GL). Compared with the 1D convolutional temporal module, in the case where the spatial component is (GL), the multi-head attention mechanism module (ATT)-based model achieves 3.15%, 1.1%, and 3.78% improvement in MAE, RMSE, and MAPE, respectively. With the spatial component being (SD), the multi-head attention mechanism module (ATT)-based model achieves 2.86%, 0.91%, and 2.59% improvement in MAE, RMSE, and MAPE, respectively.

For the task of road network prediction, previous research either started with adjacency matrix and established a static spatial matrix to represent the node relationship, or used training parameters to try to mine the node relationship in the data. In this paper, static spatial matrix and dynamic spatial matrix are constructed, respectively. The static spatial matrix is constructed by the static adjacency matrix calculated from the distance of the road network; the dynamic spatial matrix is determined by combining the historical data characteristics of the nodes and the weight matrix of the real-time self-training model. Use the static spatial matrix and the dynamic spatial matrix to perform node aggregation operations, respectively, and finally perform fusion weighting and input them into the temporal module.

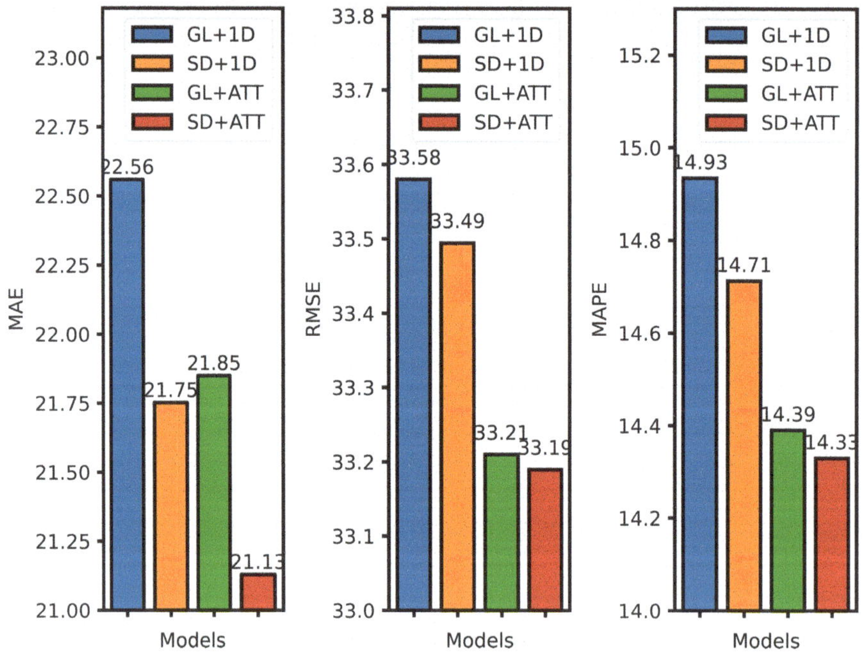

Fig. 2.9 Performance changes of different structure

Ablation experiments show that the spatial component (SD) proposed in this paper can improve the accuracy of the model.

Additionally, traffic information (vehicle flow, speed, etc.) has obvious periodicity, and the 1D convolution components used in the original model cannot extract the periodicity. Moreover, the multi-head self-attention mechanism can mine periods, and there is no disadvantage that the RNN-like model has a weak ability to capture long-term dependencies. Therefore, the multi-head attention mechanism can significantly improve the performance of the SDSCNN model.

2.3 Spatial-Temporal Complex Graph Convolution Network (ST-CGCN)

We propose a complex correlation matrix to replace the traditional adjacency matrix to better describe the relationship between different nodes. Then, we will introduce the basic structure of ST-CGCN. The structure of the ST-CGCN is shown in Fig. 2.10. To better handle the mapping relationship between different nodes, based

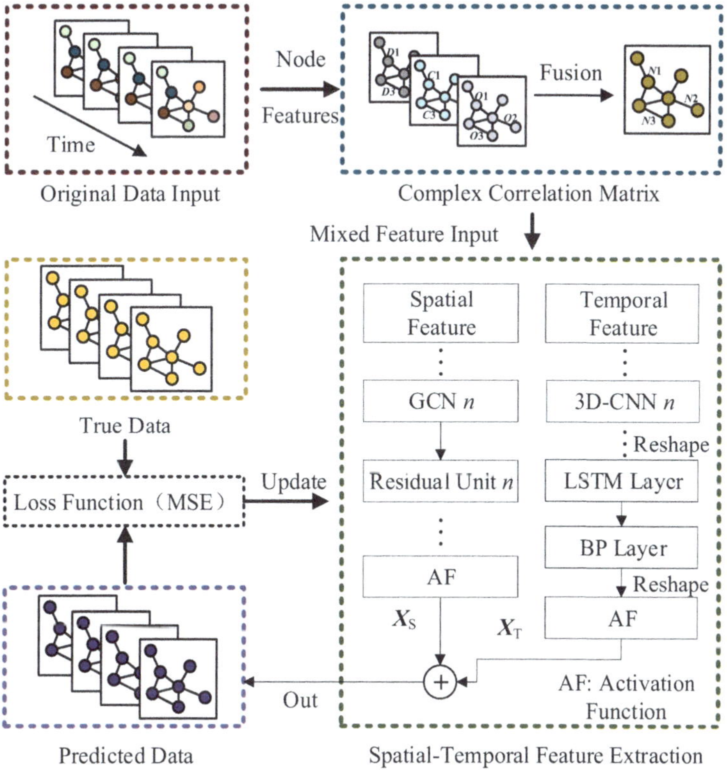

Fig. 2.10 Structure of ST-CGCN

on fully mining road location, data correlation, and weather data, we construct a distance matrix, a data correlation matrix, and a comfort measurement matrix, respectively. Then, we fuse the three matrices to obtain a complex correlation matrix. Moreover, the spatial feature extraction module uses a graph convolution operator and residual element to capture the spatial features. 3D-CNN combined with LSTM is used to build the temporal feature extraction module to capture the temporal features. The spatial and temporal features are weighted and fused to obtain the final predicted value.

2.3.1 Construction of Complex Correlation Matrix

According to the location, data correlation, and weather data of different nodes, we will construct a distance matrix, a data correlation matrix, and a comfort measurement matrix, respectively.

2.3.1.1 Distance Matrix

The distance between nodes can directly reflect the correlation of traffic flow between the longer the distance and the weaker the correlation. At present, the location information of nodes is generally based on GPS data. We collect the longitude and latitude data $\text{Point}_k = (\text{lon}_k, \text{lat}_k)$ of each node, and calculate the distance between different nodes using the Haversine formula. The Haversine formula is as follows:

$$\text{Hav}\left(\frac{d}{R}\right) = \text{Hav}\left(\text{lat}_i - \text{lat}_j\right) + \cos\left(\text{lat}_i\right)\cos\left(\text{lat}_j\right)\text{Hav}\left(\left|\text{lon}_i - \text{lon}_j\right|\right) \tag{2.14}$$

$$\text{Hav}(\theta) = \sin^2\left(\frac{\theta}{2}\right) = \frac{\left(1 - \cos(\theta)\right)}{2} \tag{2.15}$$

$$d = \theta \times R \tag{2.16}$$

where R is the radius of the earth, and the average value is 6371 km, d is the distance between the i-th node and the j-th node. Thus, the linear distance matrix $\boldsymbol{D}_1$ can be calculated.

However, the Haversine formula calculates the straight-line distance between nodes, but it is unreasonable to use the straight-line distance to measure the relationship between nodes. Some nodes have no connection but are very close to each other. To avoid the above problems, the traditional adjacency matrix is improved, and the adjacency matrix is changed into reachable matrix $\boldsymbol{D}_2$. As long as there is a connection between different nodes (either directly or indirectly), it indicates that there is a reachable relationship. The reachability relation formula is as follows:

$$\boldsymbol{D}_2(i,j)=\begin{cases}1 & N=0\\ \dfrac{1}{N} & N>0\end{cases} \tag{2.17}$$

where N is the number of nodes between two nodes, and the reachability matrix $\boldsymbol{D}_2$ of all nodes is calculated in turn.

After the calculation, two matrices about the node are obtained, which are the linear distance matrix $\boldsymbol{D}_1$ and the reachability matrix $\boldsymbol{D}_2$. The longer the distance between two nodes, the weaker their correlation. Therefore, the distance matrix $\boldsymbol{A}_{\mathrm{D}}$ can be obtained by $\dfrac{\boldsymbol{D}_2}{\boldsymbol{D}_1}$.

2.3.1.2 Data Correlation Matrix

The traffic flow data of different road nodes have data correlation. Determining how to effectively extract the correlation features between nodes is very important. We use the Pearson correlation coefficient method to calculate the correlation degree of different nodes. The Pearson correlation coefficient method is defined as:

$$\rho_{x,y}=\left[\sum_{i=1}^{m}(x_i-\bar{x})(y_i-\bar{y})\right]/\left[\sigma(x)\sigma(y)\right] \tag{2.18}$$

where x_i and y_i is the i-th feature of two nodes, respectively. $\sigma(x)$ and $\sigma(y)$ are the standard deviation of the selected feature. According to the calculation of the Pearson correlation coefficient index, the correlation between different nodes is obtained. The data correlation matrix $\boldsymbol{A}_{\mathrm{C}}$ can be obtained.

2.3.1.3 Comfort Measurement Matrix

Different weather conditions affect people's travel intentions, so a comfort measurement matrix is established to reflect people's travel intentions. We define $f_{\mathrm{comfort}}(i,j)\in[0,1]$. The better the weather in the current environment, the f_{comfort} will be close to 1. On the contrary, it will be close to 0. We collected questionnaire data from 100 people, the comfort values of people for different weather are defined as shown in Table 2.6.

We calculate the comfort value between all nodes according to $f_{\mathrm{comfort}}(i,j)\in[0,1]$ and generate the comfort measurement matrix $\boldsymbol{A}_{\mathrm{O}}$.

Table 2.6 Comfort value corresponding to different weather

Weather	Value of f_{comfort}
Sunshine fine	1
Cloudy	0.8
Overcast sky	0.7
Sprinkle	0.5
Middle rain	0.4
Drencher	0.3
Cyclone	0.1

2.3.1.4 Complex Correlation Matrix

As described in Sect. 2.3.1, three matrices are constructed to describe the relationship between nodes, namely distance matrix $\boldsymbol{A}_{\mathrm{D}}$, data correlation matrix $\boldsymbol{A}_{\mathrm{C}}$, and comfort measurement matrix $\boldsymbol{A}_{\mathrm{O}}$. Complex correlation matrix $\boldsymbol{A}_{\mathrm{N}}$ is obtained by fusing the above three matrices:

$$\boldsymbol{A}_{\mathrm{N}} = \left(\left(W_{\mathrm{D}}{}^{*}\boldsymbol{A}_{\mathrm{D}}\right) \quad \left(W_{\mathrm{C}}{}^{*}\boldsymbol{A}_{\mathrm{C}}\right) \quad \left(W_{\mathrm{O}}{}^{*}\boldsymbol{A}_{\mathrm{O}}\right)\right)\boldsymbol{W}_{\mathrm{N}} + \boldsymbol{b}_{\mathrm{N}} \tag{2.19}$$

where $\boldsymbol{W}_{\mathrm{D}}$, $\boldsymbol{W}_{\mathrm{C}}$, $\boldsymbol{W}_{\mathrm{O}}$ are the learnable weight values of the corresponding matrix, respectively. || is the concatenation operator. $\boldsymbol{W}_{\mathrm{N}}$ and $\boldsymbol{b}_{\mathrm{N}}$ are convolution output weights and bias terms of complex correlation matrix. Compared with the adjacency matrix, the complex correlation matrix considers the factors such as node distance, data association, and external interference and effectively extracts the complex correlation features of nodes, which is helpful to improve the accuracy of the traffic flow prediction model.

2.3.2 *Dynamic Spatial Feature Extraction Module*

To ensure that the proposed model has good prediction performance under the general road network structure, we use the graph convolution to extract the spatial features. The structure of the dynamic spatial feature extraction module is shown in Fig. 2.11.

Traditional GCN uses adjacency matrix $\boldsymbol{A}$ to extract the relationship between different nodes. In Sect. 2.3.1, we get a new relationship matrix $\boldsymbol{A}_{\mathrm{N}}$. The graph convolution operation is defined as:

$$\boldsymbol{X}^{l+1} = f_{\mathrm{AF}}\left(\boldsymbol{X}^{l}, \boldsymbol{A}_{\mathrm{N}}\right) = f_{\mathrm{AF}}\left(\bar{\boldsymbol{D}}^{-0.5}\bar{\boldsymbol{A}}_{\mathrm{N}}\bar{\boldsymbol{D}}^{-0.5}\boldsymbol{X}^{l}\boldsymbol{W}^{l}\right) \tag{2.20}$$

where $\boldsymbol{X}^{l}$ and $\boldsymbol{X}^{l+1}$ are the road eigenvalues before and after the extraction of the graph convolution operator. $\bar{\boldsymbol{A}}_{\mathrm{N}} = \boldsymbol{A}_{\mathrm{N}} + \boldsymbol{I}$, $\bar{\boldsymbol{D}} = \bar{\boldsymbol{D}} + \boldsymbol{I}$, $\boldsymbol{D}$ are the degree matrix of the road network. $\boldsymbol{I}$ is the identity matrix, and f_{AF} is the activation function.

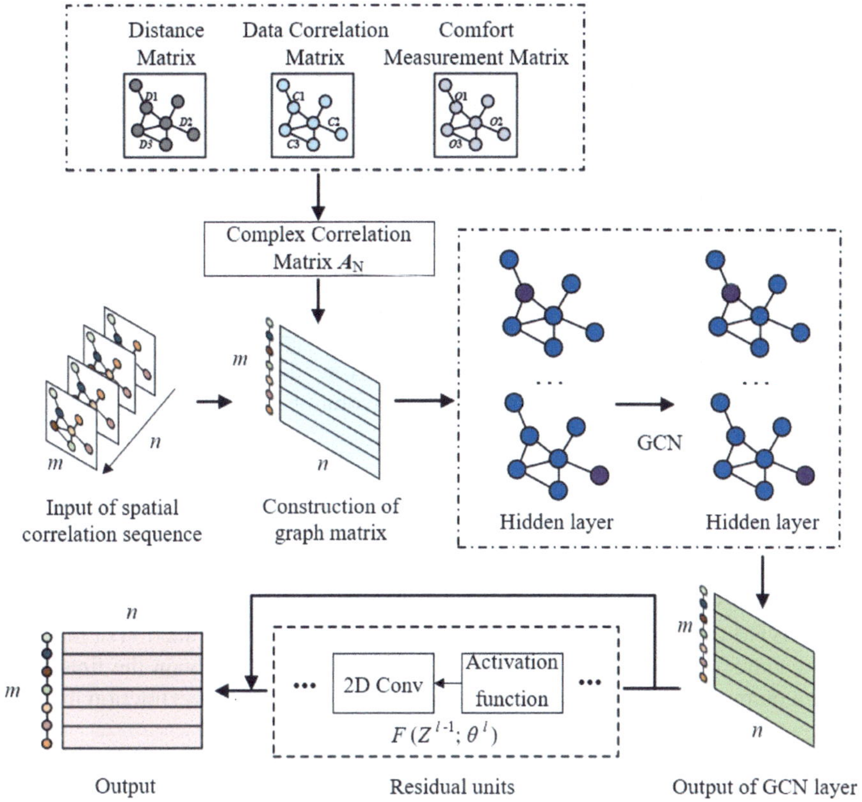

Fig. 2.11 The structure of the dynamic spatial feature extraction module

After obtaining the spatial features by using the graph convolution operator, to better extract the local small changes of the matrix, we introduce the residual element, which is defined as:

$$\boldsymbol{R}^{l+1} = \boldsymbol{R}^{l} + F\left(\boldsymbol{R}^{l}, \boldsymbol{Z}\right) \tag{2.21}$$

where $\boldsymbol{R}^{l+1}$ and $\boldsymbol{R}^{l}$ are the output and input of the residual unit, respectively. $\boldsymbol{Z}$ is the set of learnable parameters in the residual unit. F is the residual mapping. The output activation function of the spatial feature extraction module is set as the ReLU function to obtain the final spatial feature output $\boldsymbol{X}_S$.

2.3.3 Dynamic Temporal Feature Extraction Module

Traditional GCN uses convolution to extract time features, which is easy to lose the dynamic long-term temporal features of traffic data. To effectively improve the extraction of dynamic temporal features of traffic data, we introduce 3D convolution operation based on 2D convolution operation. The 3D convolution is defined as:

$$X_{t1}^{l+1} = f_{\mathrm{AF}}\left(X_{t1}^{l} W_{t1}^{l}\right) \tag{2.22}$$

where X_{t1}^{l+1} and X_{t1}^{l} are the output and input of 3D convolution, respectively, and W_{t1}^{l} is the parameter to be learned.

Although 3D convolution effectively improves the dynamic feature extraction ability of time dimension compared with 2D convolution, compared with the time series model, the time dimension cycle of 3D convolution extraction is shorter and lacks effective long-term feature extraction ability. To grasp the important time feature ability, LSTM is introduced as a long-term time dimension feature extraction module to make up for the deficiency of 3D convolution. The LSTM forward calculation operation is defined as:

$$X_{t2}^{l+1} = f_{\mathrm{LSTM}}\left(X_{t2}^{l}, W_{t2}^{l}\right) \tag{2.23}$$

where X_{t2}^{l+1} and X_{t2}^{l} are the output and input of LSTM forward unit, respectively. f_{LSTM} is the LSTM forward calculation process. W_{t2}^{l} is the parameter to be learned in the LSTM forward calculation process. The output activation function of the time feature extraction module is designed as sigmoid function to obtain the final time feature output X_{T}. The structure of the dynamic temporal feature extraction module is shown in Fig. 2.12.

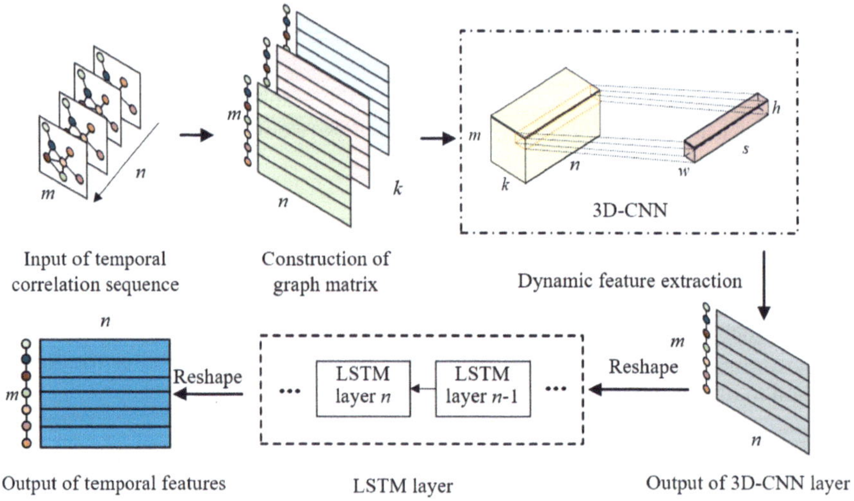

Fig. 2.12 The structure of the dynamic temporal feature extraction module

2.3.4 *Fusion and Prediction Module*

The outputs X_S and X_T of the spatial and temporal feature extraction modules are weighted and fused, and the operation is defined as:

$$X_{\text{Fusion}} = f_{\text{AF}}\left(W_1 \times X_S + W_2 \times X_T\right) \tag{2.24}$$

where W_1 and W_2 are the proportion of spatial features and temporal features, respectively.

Our ST-CGCN uses mean square error (MSE) to evaluate the errors of true data and predicted data in model training.

$$\text{MSE} = \left(\sum_{i=1}^{p}\left|X_R(i) - X_P(i)\right|^2\right)/n \tag{2.25}$$

where $X_R(i)$ and $X_P(i)$ are the real value and predicted value of the i-th sample, respectively.

2.3.5 *Experiments*

2.3.5.1 Datasets

We collect two public datasets, including PeMS03 and PeMS07, to verify the prediction performance of the proposed model. The specific information of the two datasets is shown in Table 2.7. The above two datasets are from the California Transportation Agency Performance Measurement System (PeMS), and the interval of data recording is 5 min.

To calculate the comfort measurement matrix, we need to know the weather conditions in California, so we use "https://www.almanac.com/weather/history/CA" to obtain weather data. After data sorting, traffic data shall be standardized to reduce the impact of different dimensions of data. The calculation formula is as follows:

$$X_{\text{nor}} = \left(X_{\text{real}} - \bar{X}_{\text{real}}\right)/\sigma\left(X_{\text{real}}\right) \tag{2.26}$$

Table 2.7 Information of datasets

Datasets	PeMS03	PeMS07
Location	In California, USA	In California, USA
Date	9/1/2018–11/30/2018	5/1/2017–8/31/2017
Time interval	5 min	5 min
Data type	Traffic flow	Traffic flow
Number of available time intervals	26,208	28,224

where $\boldsymbol{X}_{\text{real}}$ and $\boldsymbol{X}_{\text{nor}}$ are the data before and after standardization, respectively, $\overline{\boldsymbol{X}}_{\text{real}}$ is the average value of the data, and $\sigma(\boldsymbol{X}_{\text{real}})$ is the standard deviation of the data.

2.3.5.2 Example Calculation of Complex Correlation Matrix

In Sect. 2.3.1, we build a complex correlation matrix. In this section, we take some nodes of PeMS03 as an example to introduce the calculation methods of three matrices, respectively. We collected the longitude and latitude data of 10 sensors in VDS from 313344 to 317752, and calculate the distance and its corresponding distance matrix by using the formula in Sect. 2.3.1, as shown in Fig. 2.13a. We use formula 5 to calculate the correlation of different nodes, as shown in Fig. 2.13b. According to the location of the sensor, obtain the local historical weather data and convert it into the comfort measurement value accordingly. The weather data of the location of sensor 313344 is shown in Fig. 2.13c. Also, we can have three interesting observations. First, as can be seen in Fig. 2.13a, the farther the distance between

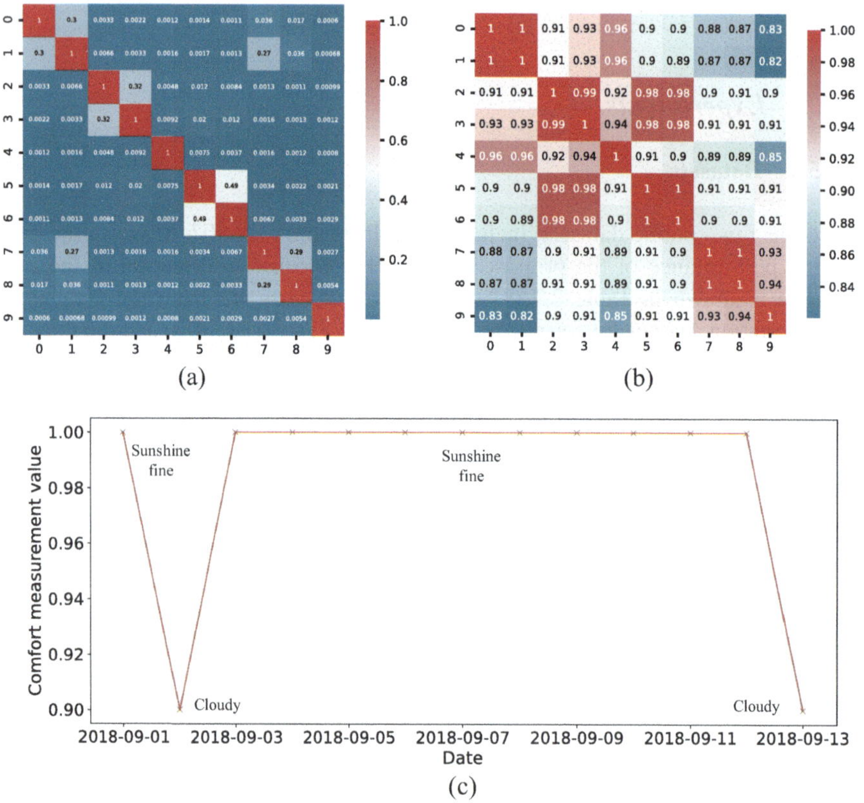

Fig. 2.13 Example calculation of complex correlation matrix in PeMS03. (a) Distance matrix. (b) Data correation matrix. (c) Comfort measurement value

nodes, the weaker the correlation between nodes. Second, it can be seen from Fig. 2.13b that the data of adjacent nodes are highly correlated. Third, it can be seen from Fig. 2.13c that different weather will affect people's comfort. To sum up, it is very important to consider the factors such as the distance, correlation, and external interference of nodes.

2.3.5.3 Experimental Settings

Hyper-Parameter

The proposed ST-CGCN is developed based on the deep learning framework Pytorch-GPU, and the graphics card is Nvidia 3060. Adam optimizer is used to optimize model parameters, and the optimization function is MSE. The training step is set 0.0001, the number of batches is set 10, and the maximum number of iterations is set 500. Other structural parameters are shown in Table 2.8.

Baseline Models and State-of-the-Art Models

We compared our proposed method with the following six models to verify the superiority of the proposed model:

- LSTM: As an improved RNN, LSTM effectively reduces the problems of gradient disappearance and gradient explosion. LSTM is usually used to extract the temporal features of traffic flow.
- CNN A typical model for extracting spatial features of Euclidean structure data.
- DCRNN: Diffusion graph convolutional network and seq2seq are introduced into RNN to extract spatial-temporal features.
- STGCN: The STGCN model uses graph convolution operation and 2D convolution network to capture spatial correlation and temporal correlation respectively.
- STSGCN: The STSGCN model designs a spatial-temporal synchronization modeling mechanism to capture complex local spatial-temporal correlations.

Table 2.8 Structure of the ST-CGCN

Parameter	ST-CGCN
Input dimension	[Batch_size, Number of nodes, 8]
Output dimension	[Batch_size, Number of nodes, 1]
Graph convolution layers	1
Dimension of 2D Conv kernel	[1×5, 1×3, 1×2]
Dimension of 3D Conv kernel	[$3 \times 1 \times 1$]
Dimension of LSTM input/output layer	[$1 \times$ Number of nodes$\times 8$]
Number of residual units	8
Activation function	Sigmoid, ReLU, Tanh

Table 2.9 Parameter settings

Parameter	LSTM	CNN, DCRNN, STGCN, STSGCN, ETGCN
Input dimension	[1,8 × Number of nodes]	[8, Number of nodes]
Output dimension	[1,1 × Number of nodes]	[1, Number of nodes]
Training times	800	800
Batch size	10	10
Training step	0.0001	0.0001
Time interval	5 min	5 min

- ETGCN [8]: The Evolution Temporal Graph Convolutional Network (ETGCN) uses similarity-based fusion method and graph convolution network to fuse multiple graph structures to learn spatial correlation.

To verify the prediction performance of ST-CGCN on traffic flow, we evaluate the proposed model by comparing the results of two classical baseline models and three state-of-the-art models. The six models are LSTM, CNN, DCRNN, STGCN, STSGCN, and ETGCN. Some parameters in all models are set uniformly to fairly verify the prediction effect. Parameter settings are shown in Table 2.9.

Evaluating Indicator

In the experiment, we use root mean square error (RMSE), mean absolute error (MAE), and mean absolute percentage error (MAPE) to scientifically evaluate the prediction effects of different models. These three formulas are as follows:

$$\mathrm{RMSE}=\sqrt{\frac{1}{p}\sum_{i=1}^{p}\left(\boldsymbol{X}_{\mathrm{R}}(i)-\boldsymbol{X}_{\mathrm{P}}(i)\right)^{2}} \tag{2.27}$$

$$\mathrm{MAE}=\frac{1}{p}\sum_{i=1}^{p}\#\boldsymbol{X}_{\mathrm{R}}(i)-\boldsymbol{X}_{\mathrm{P}}(i)\# \tag{2.28}$$

$$\mathrm{MAPE}=\frac{100\%}{p}\sum_{i=1}^{p}\left|\frac{\boldsymbol{X}_{\mathrm{R}}(i)-\boldsymbol{X}_{\mathrm{P}}(i)}{\boldsymbol{X}_{\mathrm{P}}(i)}\right| \tag{2.29}$$

2.3.5.4 Experimental Results and Discussion

Prediction Effect Based on the ST-CGCN

The ST-CGCN model is trained using the two datasets in Sect. 2.3.5, and the prediction performance of the proposed model is demonstrated using the test set. Figure 2.14a, b shows the prediction effects of ST-CGCN in the test sets of PeMS03

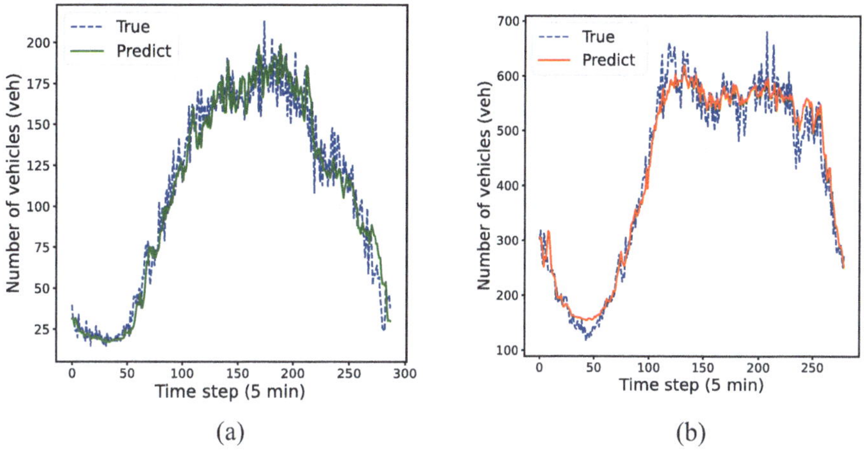

Fig. 2.14 Prediction effect based on the ST-CGCN. (**a**) Effect in the test set of PeMS03. (**b**) Effect in the test set of PeMS07

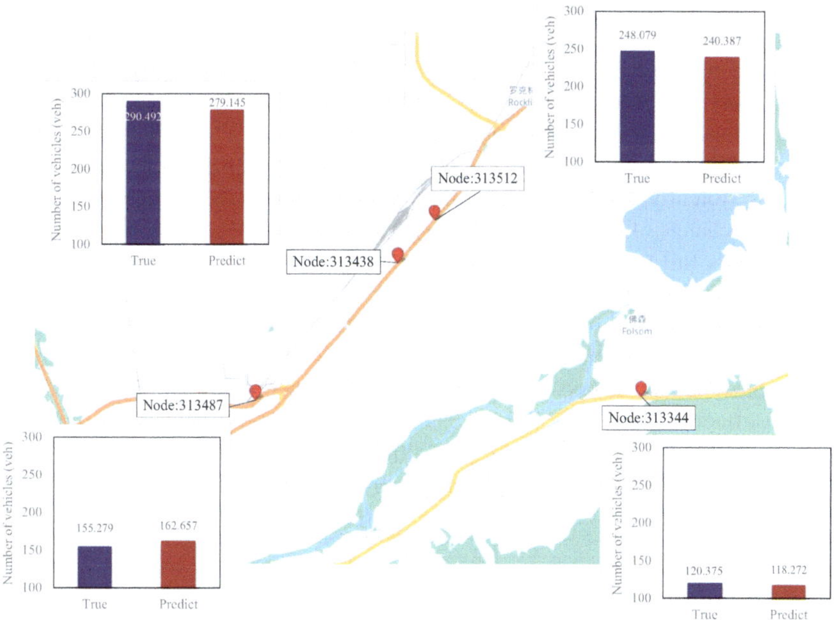

Fig. 2.15 Influence of node distribution on traffic flow

and PeMS07, respectively. As can be seen from Fig. 2.15, the traffic flow of PeMS03 reaches the maximum at noon. The traffic flow of PeMS07 has been at the peak from 12:00 to 23:00. Although the traffic flow fluctuation of the two datasets is large, the models we proposed fit the changing trend of traffic flow well.

To better show the impact of sensor distribution on traffic flow, we selected four sensors (3113344, 313438, 313487, and 313512) in PeMS03, respectively, and counted the average value of traffic flow of these four sensors for 10 days, as shown in Fig. 2.15. It can be found in Fig. 2.15 that node 313344 is not on the same road with the other three nodes, so the traffic flow of this node has a low correlation with other nodes. Since the distance between node 313438 and node 313512 is closer, the correlation between the two nodes is higher. Because the proposed complex correlation matrix takes into account the distance correlation between nodes, it can be found that our prediction results are consistent with the actual values.

Comparison with Five Classical Models

Five classical models (LSTM, CNN, DCRNN, STGCN, and STSGCN) are established according to Sect. 2.3.5 and trained using the datasets of PeMS03 and PeMS07. The average prediction results of the test set are shown in Table 2.10. The ST-CGCN proposed by us has better prediction performance. Compared with the best performance in classical models, the prediction accuracy of the proposed model is improved by 7.5%, 6.8%, and 7.5% with RMSE, MAE, and MAPE as indicators respectively.

To further compare the performance of ST-CGCN with other classical spatial-temporal feature extraction models (CNN, DCRNN, STGCN, and STSGCN), we use the test set of PeMS03 for experimental comparison, and the results are shown in Fig. 2.16. When the morning peak comes (6:00–8:00 a.m.) and the vehicle flow fluctuates violently, compared with other baseline models, ST-CGCN can capture the sudden change of flow faster. Different from the baseline models, the ST-CGCN model makes full use of the complex correlation matrix to mine the relationship between nodes and effectively improve the prediction accuracy.

As the traffic conditions change at any time of the day, we evaluated the experimental results in different periods on the dataset of PeMS03 to further study the practicability of our method. As shown in Fig. 2.17, the proposed model can achieve the best effect in most periods, especially in the periods when the traffic speed changes sharply, such as 10:00–12:00, 12:00–14:00, 18:00–20:00, and 20:00–22:00. Due to the ST-CGCN model effectively extracting the dynamic spatial-temporal

Table 2.10 Performance comparison of different models in the test set (Average)

	PeMS03			PeMS07		
Model	RMSE	MAE	MAPE (%)	RMSE	MAE	MAPE (%)
LSTM	35.233	21.132	25.561	45.927	30.021	13.532
CNN	32.987	20.725	20.773	40.794	32.141	12.361
DCRNN	30.427	18.239	18.924	38.856	25.764	11.693
STGCN	30.314	17.485	17.472	38.758	25.382	11.087
STSGCN	28.873	17.379	16.882	38.894	24.396	10.213
ST-CGCN	**25.714**	**15.665**	**15.175**	**37.281**	**23.491**	**9.709**

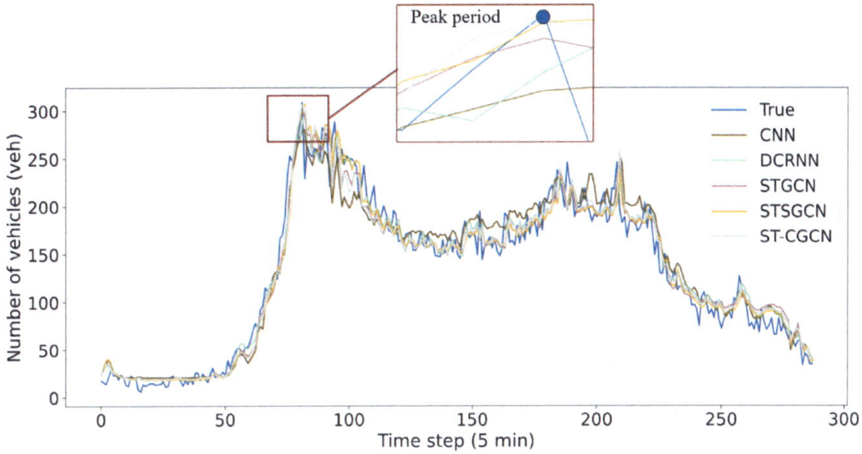

Fig. 2.16 Comparison of 1-day prediction curve in the test set of PeMS03

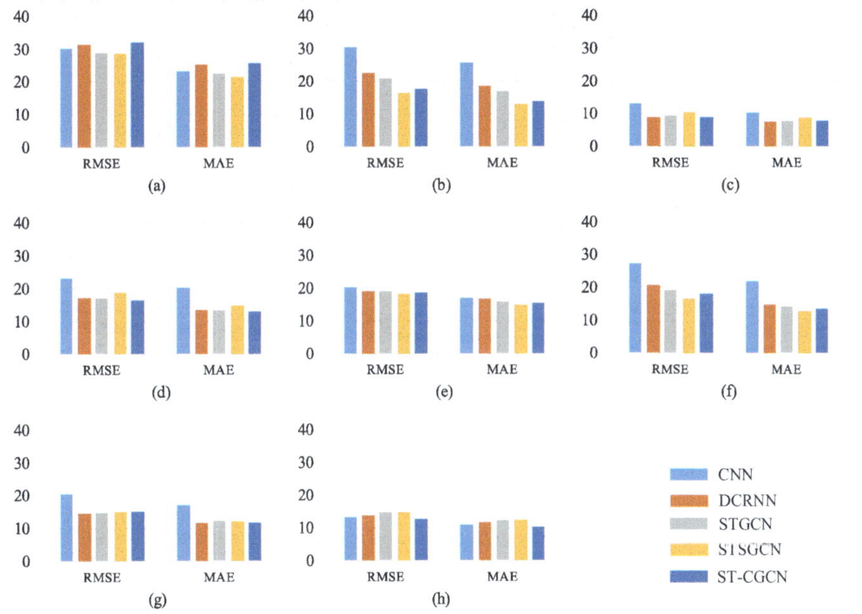

Fig. 2.17 Performance comparison of different methods during different periods on the dataset of PeMS03. (**a**) 6:00–8:00. (**b**) 8:00–10:00. (**c**) 10:00–12:00. (**d**) 12:00–14:00. (**e**) 14:00–16:00. (**f**) 16:00–18:00. (**g**) 18:00–20:00. (**h**) 20:00–22:00

features of traffic flow based on mining node association features, the proposed model has better prediction accuracy and inflection point capture ability than the state-of-the-art models.

In order to improve the long-term time feature capture ability, we introduce 3D convolution and LSTM components into the proposed model. The dataset of PeMS03 is used to verify the long-term feature capture ability. RMSE, MAE, and MAPE are used as evaluation indicators. The comparison results with other baseline models are shown in Fig. 2.18. The prediction interval is from 5 to 60 min. The experimental results show that compared with other models, our proposed model can always maintain high-prediction accuracy with the increase in prediction interval.

Comparison with the Model Considering Multi-Graph Adjacency Matrix

The ETGCN model also optimizes the adjacency matrix. Different from ETGCN, the ST-CGCN model proposed by us effectively integrates the external interference and effectively captures the dynamic temporal features of traffic flow by using 3D convolution operation and LSTM. Because some parameters in ETGCN are not clear, in this experiment, we only use the combination of three improved adjacency matrices of ETGCN and GCN as the test model. Two datasets (PeMS03 and PeMS07) are used to test the prediction accuracy of the two models, and RMSE is used as the evaluation index. The experimental results as shown in Table 2.11 show that the proposed model has better prediction performance.

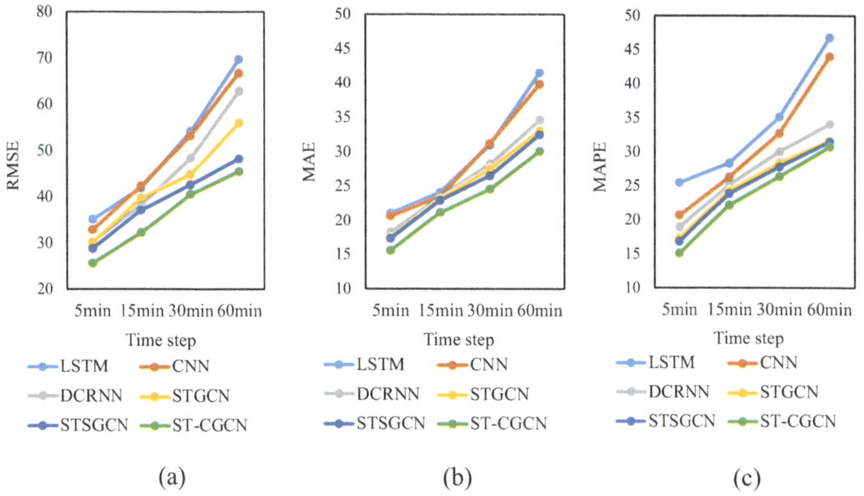

Fig. 2.18 Comparison of multiple prediction results (5–60 min). (**a**) RMSE. (**b**) MAE. (**c**) MAPE

Table 2.11 Comparison with the model considering multi-graph adjacency matrix

	PeMS03	PeMS07
Model	RMSE	RMSE
ETGCN	26.374	38.022
ST-CGCN	25.714	37.281

Discussion

In past research, LSTM and CNN are often used to extract spatial-temporal features. Because LSTM focuses on the extraction of temporal features, LSTM is easy to lose the spatial features of traffic flow, resulting in low prediction accuracy. When CNN processes Euclidean structure data, it transforms the road network into a regular grid structure. CNN's prediction accuracy of grid area is high, but the non-Euclidean structure data is irregular graphics, which makes CNN unable to effectively predict arbitrarily distributed sensor nodes. Different from LSTM and CNN, DCRNN, STGCN, STSGCN can effectively extract the spatial-temporal features through the combination of multiple components, but they only use adjacency matrix to describe the association features of nodes, which cannot effectively reflect the factors: node distance, data correlation, and external interference. The ETGCN model optimizes the adjacency matrix, but it ignores the interference of external factors. The proposed model uses a complex correlation matrix to describe the correlation features of sensor nodes, takes distance, data correlation and external interference as important considerations, uses graph convolution and residual unit to capture the dynamic spatial features, and uses 3D convolution and LSTM to capture the dynamic temporal features. The results in Sect. 2.3.5 prove that our model has advantages in multi-step prediction (5–60 min) and can effectively capture dynamic spatial-temporal features. Compared with the best performance in state-of-the-art models, the prediction accuracy of the proposed model is improved by 2.2% with RMSE as indicators, respectively. Therefore, the proposed model has better prediction performance in a complex traffic environment.

2.4 Prior Knowledge Enhanced Time-Varying Graph Convolution Network (PKET-GCN)

Figure 2.19 shows the structure of PKET-GCN. A prior knowledge graph is initially constructed to extract the sensor node's complex spatial-temporal correlation features. Next, a novel dynamic spatial and time-varying module is constructed. Finally, a projection module is designed to map the predicted future traffic flow value.

2.4.1 Prior Knowledge Graph

Figure 2.20 shows the prior knowledge graph construction. Firstly, the static node feature matrix is generated according to the location of the sensor nodes, after which the dynamic node feature matrix is generated according to the data correlation and the environment variables of the sensor nodes. Then, the static node feature matrix and dynamic node feature matrices are fused into a priori knowledge graph using an encoder–decoder framework and self-attention.

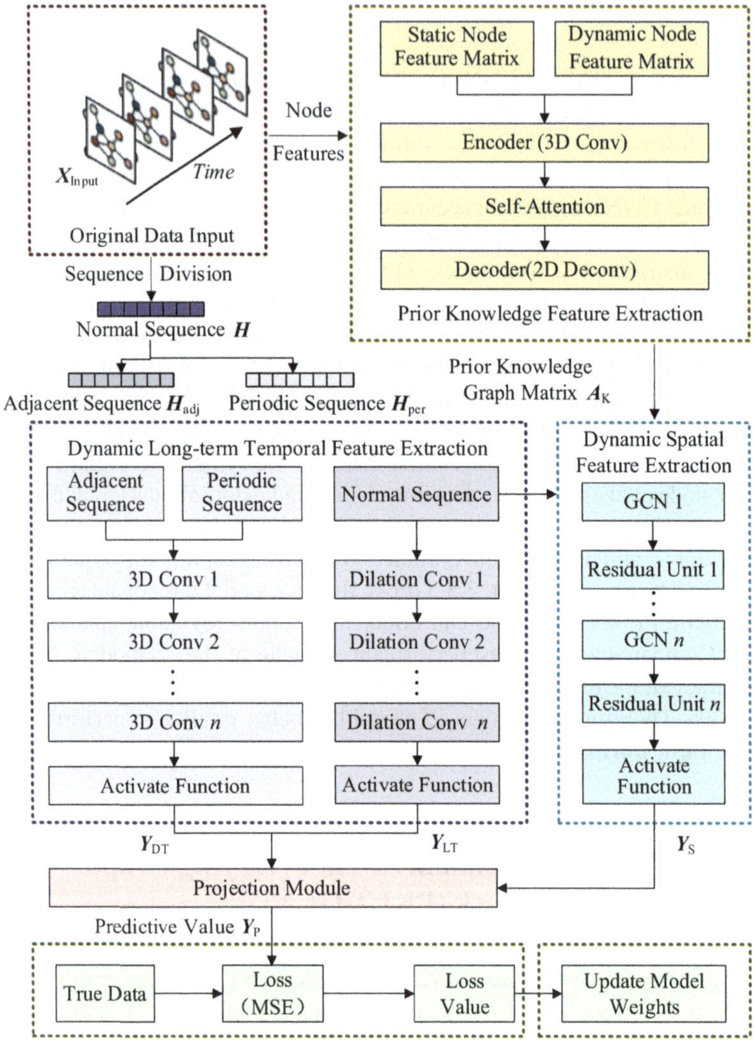

Fig. 2.19 Structure of PKET-GCN

2.4.1.1 Static Node Feature Matrix

The static node characteristic matrix is mainly calculated using the effective distance between nodes. According to the longitude and latitude $\{\text{lon}_n, \text{lat}_n\}$ of sensor n and the longitude and latitude data $\{\text{lon}_m, \text{lat}_m\}$ of sensor m, the physical distance between nodes is calculated using the Haversine formula, as described below.

$$\text{Hav}\left(\frac{d}{R}\right) = \text{Hav}\left(\text{lat}_m - \text{lat}_n\right) + \cos\left(\text{lat}_m\right)\cos\left(\text{lat}_n\right)\text{Hav}\left(\lVert\text{lon}_m - \text{lon}_n\rVert\right) \tag{2.30}$$

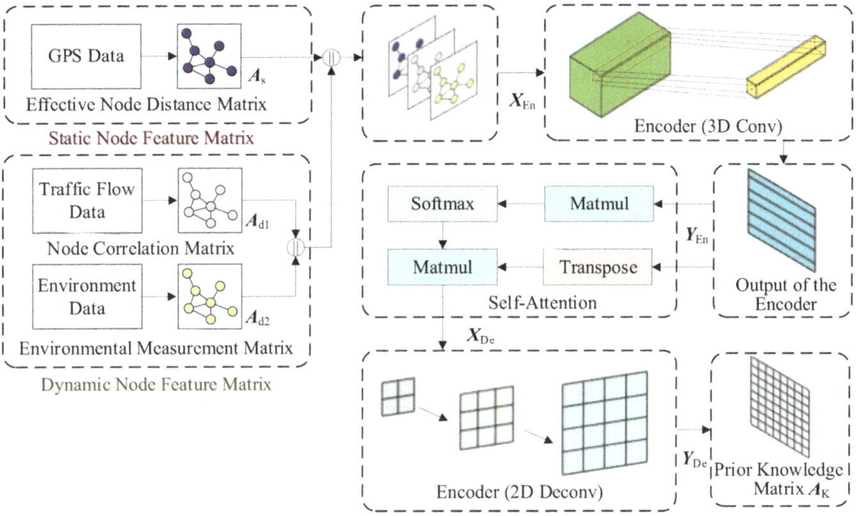

Fig. 2.20 The construction process of a prior knowledge graph

where $\mathrm{Hav}\left(\frac{d}{R}\right)=\sin^2\left(\frac{d}{2R}\right)$, R is the radius of the earth, the average value of R is 6371 km, and d is the physical distance between sensors m and n. Thus, the physical distance d_{mn} between any two sensors can be calculated.

Although some nodes have the closest linear distance, there is no direct connection between the roads. Hence, the physical distance d_{mn} cannot effectively reflect the interaction of nodes. To this end, the reachability relationship discriminant of any two nodes is established:

$$p_{mn}=\begin{cases}1 & N=0\\ N & N>0\end{cases} \tag{2.31}$$

where N is the number of sensors between two nodes. Physical distance d_{mn} and reachability relation discriminant p_{mn} are then combined to generate the effective distance $\mathrm{Ed}_{mn}=\frac{1}{e^{p_{mn}\times d_{mn}}}$. The effective node distance matrix is generated according to the effective distance between any two nodes. Therefore, the static node feature matrix is $\boldsymbol{A}_s$.

2.4.1.2 Dynamic Node Feature Matrix

Dynamic node features are constructed to extract the dynamic association between nodes. The node correlation matrix and environment measurement matrix are built so as to reflect the dynamic data correlation and environmental suitability between nodes.

Node Correlation Matrix $\boldsymbol{A}_{\mathrm{d1}}$

As the traffic flow data recorded between sensor nodes had correlated data, the Pearson correlation coefficient (PCCs) is used to calculate the correlation coefficient, as given below.

$$\rho_{mn} = \left[\sum_{t=1}^{T} \left(\boldsymbol{H}_m(t) - \mathrm{Mean}(\boldsymbol{H}_m) \right) \left(\boldsymbol{H}_n(t) - \mathrm{Mean}(\boldsymbol{H}_n) \right) \right] / \left(\mathrm{Std}(\boldsymbol{H}_m) \mathrm{Std}(\boldsymbol{H}_n) \right) \tag{2.32}$$

where $\boldsymbol{H}_m(t)$ and $\boldsymbol{H}_n(t)$ are the features of sensor m and n at time t, respectively, $\mathrm{Std}(\boldsymbol{H}_m)$ and $\mathrm{Std}(\boldsymbol{H}_n)$ are the standard deviation of the selected features, respectively, and $\mathrm{Mean}(\boldsymbol{H}_m)$ and $\mathrm{Mean}(\boldsymbol{H}_n)$ are the mean values of the selected features, respectively. The node correlation matrix $\boldsymbol{A}_{\mathrm{d1}}$ is generated according to the correlation coefficient between any two nodes.

Environmental Measurement Matrix $\boldsymbol{A}_{\mathrm{d2}}$

The environmental measurement matrix is mainly composed of environmental variables, which mainly included weather, climate, and special events. For example, in terms of weather, the better the weather, the greater the people's desire to travel. However, the worse the weather, the less their desires are. The environmental measurement e_{mn} between sensors m and n, $e_{mn} \in [0,1]$ was defined along with the environmental measurement values, as shown in Table 2.12. According to the weather data $\{\mathrm{Weather}_m\}$ and $\{\mathrm{Weather}_n\}$ at the location of sensors m and n, the environmental measurement e_{mn} is calculated between sensors using Table 2.12. The values in Table 2.12 are the quantitative results after randomly selecting 100 people for questionnaire survey. The environmental measurement matrix $\boldsymbol{A}_{\mathrm{d2}}$ is generated according to the environment measurement values between any two nodes.

Table 2.12 Environment measurement values corresponding to different types of weather

Weather	Value
Sunshine fine	1
Rain	0.5
Extreme weather	0.1

2.4.1.3 Prior Knowledge Extraction Module

It is crucial to model the joint contribution of the nodes' dynamic and static features; losing essential information by weighted fusion is also common. Considering that the encoder–decoder framework can adequately obtain information in natural language processing, the encoder–decoder is taken as the basic framework, introducing self-attention to better extract correlation features.

The encoder module is mainly stacked using a 3D convolution operation. By adopting the subsampled method, the features of the series matrix of the input static and dynamic node feature matrix are extracted with the following encoder input features:

$$\boldsymbol{X}_{\mathrm{En}} = \boldsymbol{A}_{\mathrm{s}} \quad \boldsymbol{A}_{\mathrm{d1}} \quad \boldsymbol{A}_{\mathrm{d2}} \tag{2.33}$$

where || is the concatenation operator. The operation of the encoder structure is defined as:

$$\boldsymbol{X}_{\mathrm{En}}^{l+1} = f_{\mathrm{A}}\left(f_{\mathrm{En}}\left(\boldsymbol{X}_{\mathrm{En}}^{l} * \boldsymbol{W}_{\mathrm{En}}^{l}\right)\right) \tag{2.34}$$

where f_{A} is the activation function, f_{EN} is the encoder operation, $\boldsymbol{X}_{\mathrm{En}}^{l}$ and $\boldsymbol{X}_{\mathrm{En}}^{l+1}$ are the input and output of the encoder operation, respectively, $\boldsymbol{W}_{\mathrm{En}}^{l}$ is the parameter to be learned by the encoder, and * is the convolution operation. After multiple encoding, the final output of the encoder is defined as $\boldsymbol{Y}_{\mathrm{En}}$.

Wang et al. demonstrated through experiments that the self-attention module plays a vital role in extracting the key features of traffic flow. Therefore, self-attention is added to assign more weights to key nodes after the encoder operation, defining the operation of the self-attention as:

$$\boldsymbol{X}_{\mathrm{De}} = \left(\boldsymbol{Y}_{\mathrm{En}}\right)^{\mathrm{Tr}} \cdot \mathrm{Atten}\left(\boldsymbol{Y}_{\mathrm{En}}\right) \tag{2.35}$$

$$\mathrm{Atten}\left(\boldsymbol{Y}_{\mathrm{En}}\right) = \frac{\mathrm{e}^{\left(\boldsymbol{Y}_{\mathrm{En}} \cdot \left(\boldsymbol{Y}_{\mathrm{En}}\right)^{\mathrm{Tr}}\right)}}{\sum_{1}^{T} \mathrm{e}^{\left(\boldsymbol{Y}_{\mathrm{En}} \cdot \left(\boldsymbol{Y}_{\mathrm{En}}\right)^{\mathrm{Tr}}\right)}} \tag{2.36}$$

where Tr is a transpose operation. The self-attention extracts the output of the encoder operation to obtain the input feature $\boldsymbol{X}_{\mathrm{De}}$ of the decoder structure.

Following the encoder and self-attention operations, a decoder structure is constructed to decode the encoded information. The decoder structure is mainly composed of 2D deconvolution, while the output of the self-attention feature is decoded via up-sampling, whose operation is defined as:

$$\boldsymbol{X}_{\mathrm{De}}^{l+1} = f_{\mathrm{A}}\left(f_{\mathrm{DE}}\left(\boldsymbol{X}_{\mathrm{De}}^{l} * \boldsymbol{W}_{\mathrm{De}}^{l}\right)\right) \tag{2.37}$$

where X_{De}^{l} and X_{De}^{l+1} are the input and output of the decoder operation, respectively. W_{De}^{l} is a parameter to be learned in the decoder structure, and f_{DE} is the encoder operation. After multiple decoding operations, the final output of the decoder is defined as Y_{De}.

The final output Y_{De} of multiple decoder operations is the prior knowledge matrix $A_{\mathrm{K}} \in R^{N \times N}$, which is the fusion of the static node feature matrix and dynamic node feature matrices.

2.4.2 *Dynamic Spatial Feature Extraction Module*

Unlike classical graph convolution, the adjacency matrix is not used as the correlation measure of nodes. Instead, the prior knowledge graph is utilized to model the dynamic and static features of the nodes jointly. Figure 2.21 shows the dynamic spatial feature extraction module.

To avoid the problem that the model accuracy decreases with too many convolution layers, the residual module is introduced after the graph convolution operation. Therefore, the improved graph convolution operation is defined as:

$$H^{l+1} = f_{\mathrm{A}}\left(H^{l}, A\right) = f_{\mathrm{A}}\left(A_{\mathrm{K}} H^{l} W^{l}\right) + F\left(H^{l}, Z^{l}\right) \tag{2.38}$$

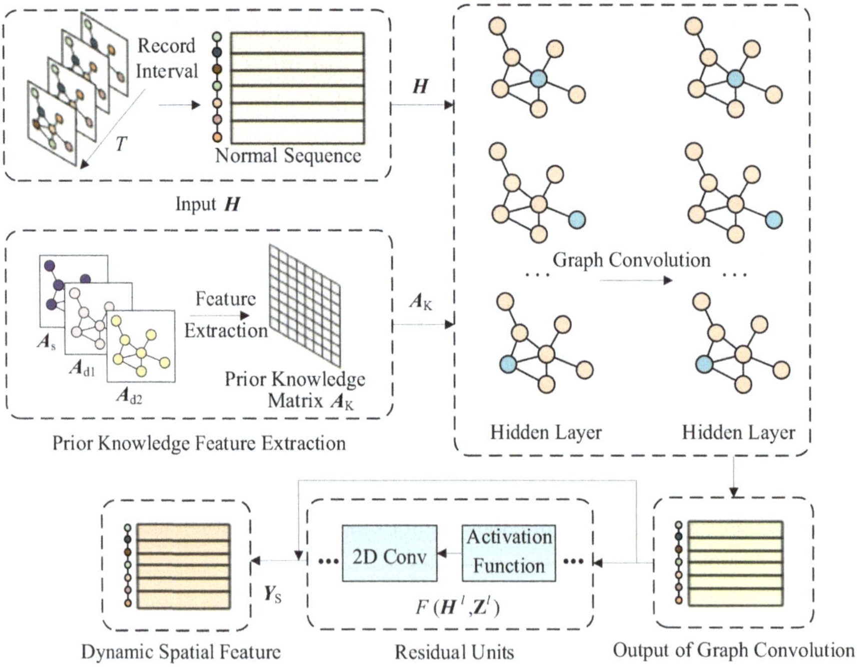

Fig. 2.21 The structure of the dynamic spatial feature extraction module

where $\boldsymbol{H}^l$ and $\boldsymbol{H}^{l+1}$ are the road feature values before and after the graph convolution, respectively. $\boldsymbol{W}^l$ and $\boldsymbol{Z}^l$ are the weights of the graph convolution and the residual operation, respectively, and F is the residual operation. The output of the dynamic spatial feature extraction module is $\boldsymbol{Y}_{\mathrm{S}}$.

2.4.3 *Time-Varying Feature Extraction Module*

To effectively extract dynamic long-term temporal features, this paper designs a time-varying feature extraction module composed of a dynamic temporal feature extraction module and long-term temporal feature extraction module. Figure 2.22 shows the structure of the time-varying feature extraction module.

2.4.3.1 Dynamic Temporal Feature Extraction Module

Traffic flow data $\boldsymbol{H}$ is divided into adjacent sequence $\boldsymbol{H}_{\mathrm{adj}}$ and periodic sequence $\boldsymbol{H}_{\mathrm{per}}$ according to different time intervals, where the adjacent sequence time interval is 1 h, and the periodic sequence time interval is 24 h. Two sequence features are then concatenated, and the dynamic temporal feature is extracted by 3D convolution operation.

The input feature is defined as:

$$\boldsymbol{X}_{\mathrm{DT}} = \boldsymbol{H}_{\mathrm{adj}} \quad \boldsymbol{H}_{\mathrm{per}} \tag{2.39}$$

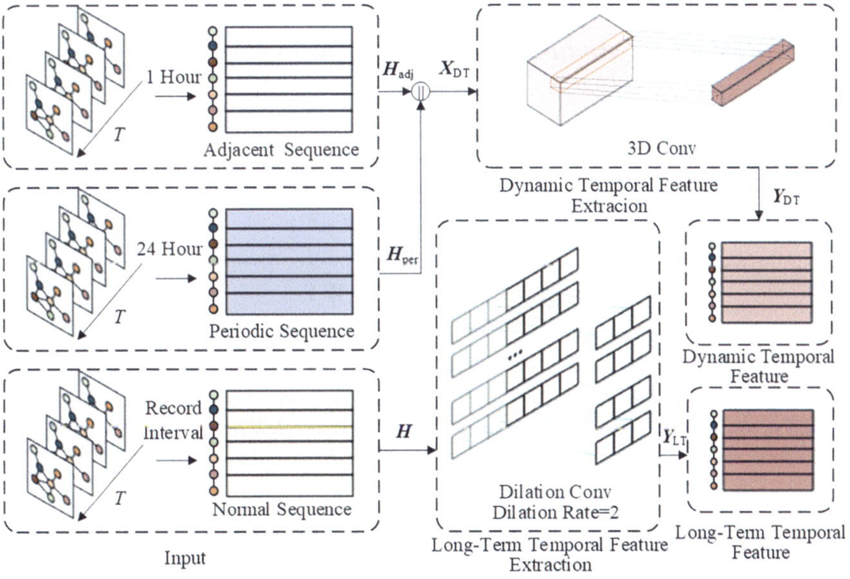

Fig. 2.22 The structure of the time-varying feature extraction module

The dynamic temporal feature extraction operation is defined as:

$$\boldsymbol{X}_{\mathrm{DT}}^{l+1} = f_{\mathrm{A}}\left(f_{\mathrm{DT}}\left(\boldsymbol{X}_{\mathrm{DT}}^{l} {}^{*}\boldsymbol{W}_{\mathrm{DT}}^{l}\right)\right) \tag{2.40}$$

where $\boldsymbol{X}_{\mathrm{DT}}^{l}$ and $\boldsymbol{X}_{\mathrm{DT}}^{l+1}$ are the input and output of the dynamic temporal feature extraction operation, respectively, $\boldsymbol{W}_{\mathrm{DT}}^{l}$ is a parameter to be learned, and f_{DT} is a dynamic feature extraction operation. The output of the dynamic temporal feature extraction module is $\boldsymbol{Y}_{\mathrm{DT}}$.

2.4.3.2 Long-Term Temporal Feature Extraction Module

As LSTM or GRU requires several iterations to extract long-term temporal features, dilation convolution in TCN is used to extract the long-term temporal features, which is defined as:

$$\boldsymbol{X}_{\mathrm{LT}}^{l+1} = f_{\mathrm{A}}\left(f_{\mathrm{LT}}\left(\boldsymbol{X}_{\mathrm{LT}}^{l} {}^{*}\boldsymbol{W}_{\mathrm{LT}}^{l}\right)\right) \tag{2.41}$$

where $\boldsymbol{X}_{\mathrm{LT}}^{l}$ and $\boldsymbol{X}_{\mathrm{LT}}^{l+1}$ are the input and output of the dilation convolution operation, respectively, $\boldsymbol{X}_{\mathrm{LT}} = \boldsymbol{H}$, and $\boldsymbol{W}_{\mathrm{LT}}^{l}$ is the parameter to be learned of dilation convolution kernel, and f_{LT} is a long-term feature extraction operation. Different from other 2D convolutions, the expansion rate of the convolution kernel in expansion convolution is set to 2 to collect a wider range of features. The output of the long-term temporal feature extraction module is $\boldsymbol{Y}_{\mathrm{LT}}$.

2.4.4 *Projection Module and Model Training*

According to the gated fusion mechanism, a projection module is designed to connect the outputs of different modules and map the final predictive value. The structure of the projection module is illustrated in Fig. 2.23. Here, the projection module is defined as:

$$\boldsymbol{Y}_{\mathrm{P}} = \boldsymbol{W}_{\mathrm{T}}\left(\boldsymbol{Y}_{\mathrm{DT}} \quad \boldsymbol{Y}_{\mathrm{LT}}\right) \circ \boldsymbol{W}_{\mathrm{S}}\boldsymbol{Y}_{\mathrm{S}} \tag{2.42}$$

where $\boldsymbol{W}_{\mathrm{T}}$ and $\boldsymbol{W}_{\mathrm{S}}$ are the weights of dynamic spatial and long-term temporal feature extraction components, respectively, and $\circ$ is the Hadamard product.

The proposed PKET-GCN uses mean square error (MSE) as a loss function:

$$\mathrm{MSE} = \left(\sum_{i=1}^{p}\left|\boldsymbol{Y}_{\mathrm{Real}}(i) - \boldsymbol{Y}_{\mathrm{Pred}}(i)\right|^{2}\right) / p \tag{2.43}$$

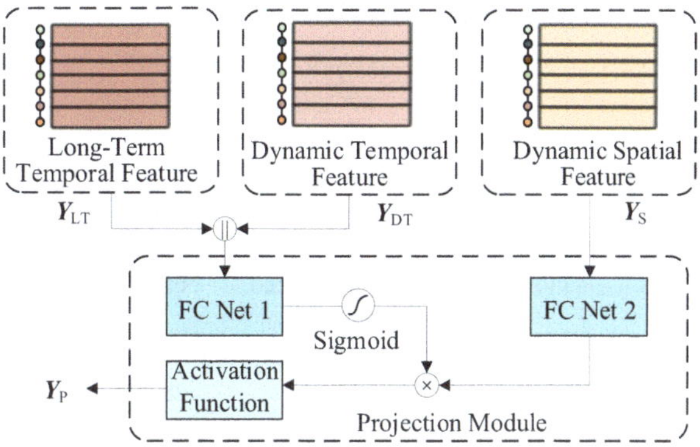

Fig. 2.23 The structure of the projection module

where $\boldsymbol{Y}_{\mathrm{R}}$ and $\boldsymbol{Y}_{\mathrm{P}}$ are the true and predictive values, respectively, and p is the number of samples.

2.4.5 *Training Algorithm of PKET-GCN*

It is very important to select an appropriate model training algorithm. Algorithm 1 shows the training process of PKET-GCN, with $\boldsymbol{X}_{\mathrm{train}}$ as the training set. Firstly, a prior knowledge matrix $\boldsymbol{A}_{\mathrm{K}}$ and three cycles $\boldsymbol{H}_{\mathrm{adj}}$, $\boldsymbol{H}_{\mathrm{per}}$, $\boldsymbol{H}$, are constructed from the original data. The proposed model is then repeatedly trained using forward and backward propagation. The Adam algorithm is chosen to optimize PKET-GCN.

Algorithm 1 PKET-GCN Training Algorithm

Input: The input traffic flow data: $\{\boldsymbol{H}_t \mid t = 1, 2, \ldots, T\}$; the prior knowledge matrix $\boldsymbol{A}_K$; the lengths of adjacent and periodic sequence are La, Lp; the predicted time interval Δt
Output: PKET-GCN with optimal weight
$\varnothing \rightarrow \mathbf{X}_{train}$
// construction of training samples
for all available time interval $t \in [1, T]$ **do**
$\boldsymbol{H}_{adj} = [H_{t-La}, H_{t-(La-1hour)}, \ldots, H_{t-1hour}]$
$\boldsymbol{H}_{per} = [\mathrm{H}_{t-Lp}, \mathrm{H}_{t-(La-24hour)}, \ldots, \mathrm{II}_{t-24hour}]$
$//H_{T+\Delta t}$ is the target value to be predicted
put a sample $\{\boldsymbol{H}_t, \boldsymbol{H}_{adj}, \boldsymbol{H}_{per}, \boldsymbol{A}_K\}$ into $\boldsymbol{X}_{train}$
end for
// train PKET-GCN
initialize all parameters of PKET-GCN
repeat
randomly select a batch of samples $\boldsymbol{X}_{min\_batch}$ from $\boldsymbol{X}_{train}$
find the optimal parameters of $\boldsymbol{X}_{min\_batch}$ to optimize the model
until end if iteration conditions are met
output PKET-GCN with optimal weight

Table 2.13 Information of datasets

Datasets	PeMS03	PeMS04	PeMS07	PeMS08	SZ-taxi
Location	In California, USA	In California, USA	In California, USA	In California, USA	In Shenzhen, China
Date	9/1–11/30/2018	1/1–2/28/2018	5/1–8/31/2017	7/1–8/31/2016	1/1–1/31/2015
Time interval	5 min	5 min	5 min	5 min	15 min
Data type	Flow	Flow, speed, occupancy	Flow	Flow, speed, occupancy	Speed
Nodes	358	307	883	170	156
Intervals	26,208	16,992	28,224	17,856	2976

2.4.6 Experiments

2.4.6.1 Real-World Datasets

This study collected five public datasets, including PeMS03, PeMS04, PeMS07, PeMS08, and SZ-taxi. The specific information pertaining to the five real-world datasets is shown in Table 2.13. Four PeMS datasets were proposed by the California Transportation Agency Performance Measurement System (PeMS). The SZ-taxi dataset was collected in Luohu District, Shenzhen, China, and the traffic speeds at different times on the road were recorded.

To calculate the prior knowledge graph, GPS and weather data corresponding to each dataset were collected. Weather data were obtained from "https://www.almanac.com/weather/history/CA," and the GPS data of the PeMS datasets were obtained from "https://pems.dot.ca.gov/." Since the node distribution of the SZ-taxi dataset was unknown, the adjacency matrix was used to represent the distance relationship between nodes. After data collection, the data were standardized. The specific standardization formula used is shown below.

$$\boldsymbol{X}_{\text{nor}} = \left(\boldsymbol{X}_{\text{real}} - \text{Mean}\left(\boldsymbol{X}_{\text{real}}\right)\right) / \text{Std}\left(\boldsymbol{X}_{\text{real}}\right) \tag{2.44}$$

where $\boldsymbol{X}_{\text{nor}}$ and $\boldsymbol{X}_{\text{real}}$ are the data after and before standardization, respectively.

Next, the determination process of static and dynamic node features is outlined. Figure 2.24 shows the static node features matrix of the first ten sensors in PeMS03. According to Fig. 2.24, the distance between the sensors with ID = 0 and ID = 1 is relatively close; hence, the static node feature values of the two sensors are high. Moreover, the distance between sensors with ID = 2 and ID = 1 is relatively long, resulting in small static node eigenvalues. Figure 2.25 shows the node correlation matrix of the top ten sensors from PeMS03 to PeMS08. In regard to PeMS03, the sensors with ID = 1 and ID = 2 are shown to be close and have high data correlation between them as a result. The weather data of the location of the SZ-taxi dataset are then collected to demonstrate the calculation of the environmental measurement matrix. Figure 2.26a shows the weather type statistics of the SZ-taxi dataset, and

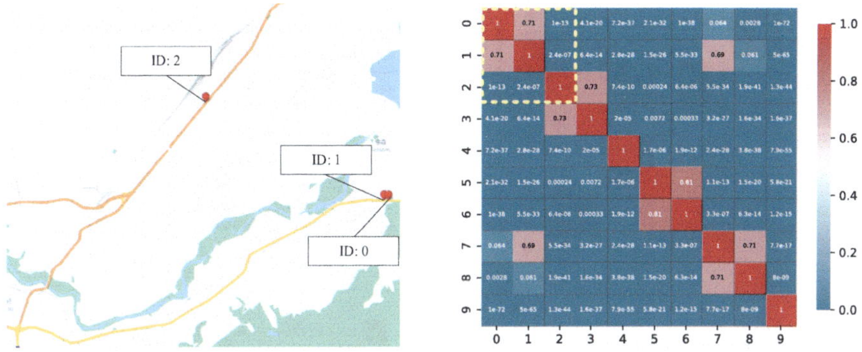

(a) (b)

Fig. 2.24 Calculation diagram of effective node distance matrix. (**a**) Distribution of nodes. (**b**) Effective node distance matrix

Fig. 2.26b shows the corresponding environmental measurement values of the SZ-taxi dataset.

2.4.6.2 Experimental Settings

Hyper-Parameter

This study utilized a system running on NVIDIA GeForce RTX 3080 GPU, Intel Core i7-12700 CPU (2.1 GHz), 32 GB RAM, and Windows 11 64-bit in conducting the experiments using PyTorch-GPU. The Adam optimizer was used to optimize the network parameters, while the loss function was MSE. The initial learning rate was set to 0.0001, the number of batches was set to 30, and the maximum number of iterations was set to 200. The five datasets were then divided into training sets, and test sets at a ratio of 7:3. Other structural parameters of the PKET-GCN model are shown in Table 2.14.

Baseline Models and State-of-the Art Models

Two baseline models and four state-of-the-art models were established to compare the prediction performance with the PKET-GCN model.

- LSTM: As an improved model of RNN, The LSTM model can effectively reduce the disappearance of the frontal gradient. LSTM is often used to extract temporal features.
- CNN: The CNN model uses a regular convolution kernel to extract spatial features of Euclidean structure.
- DCRNN: The DCRNN model uses a two-way random walk to model spatial dependence and encoder–decoder architecture to model temporal dependence.

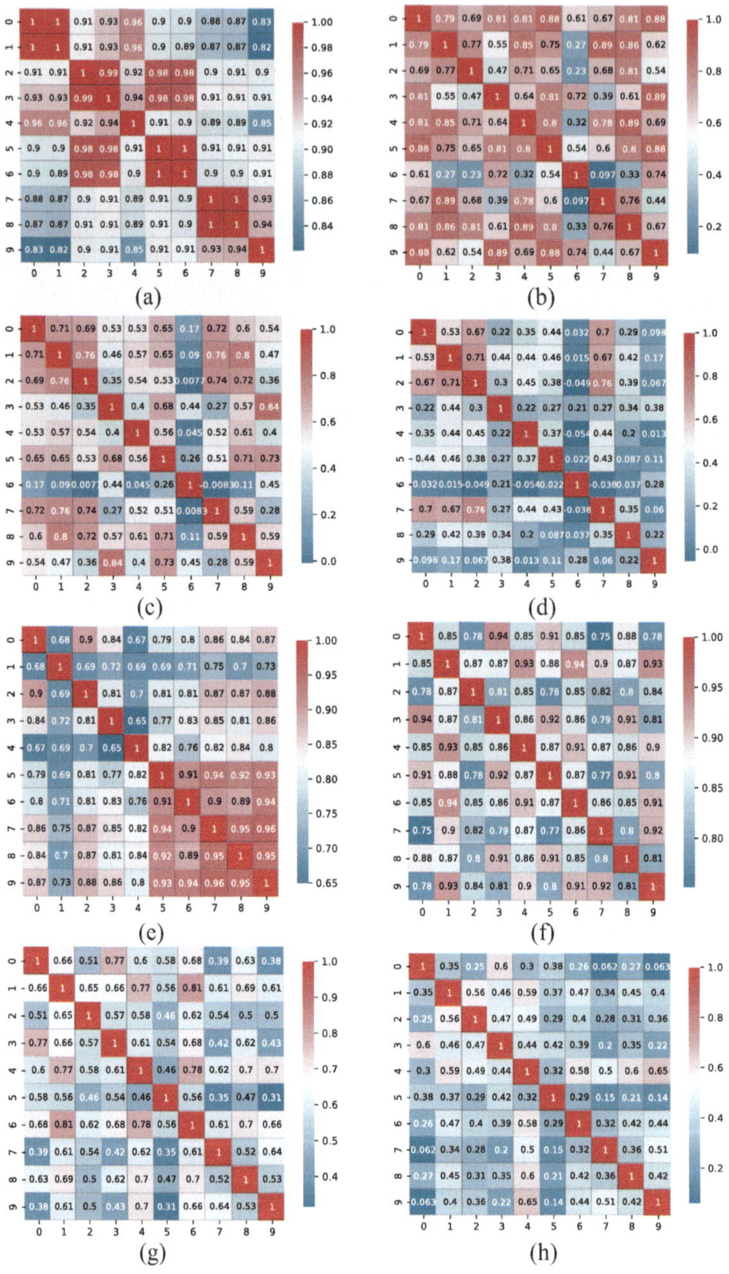

Fig. 2.25 Calculation diagram of the node correlation matrix. (**a**) PeMS03 (flow). (**b**) PeMS04 (flow). (**c**) PeMS04 (occupancy). (**d**) PeMS04 (speed). (**e**) PeMS07 (flow). (**f**) PeMS08 (flow). (**g**) PeMS08 (occupancy). (**h**) PeMS08 (speed)

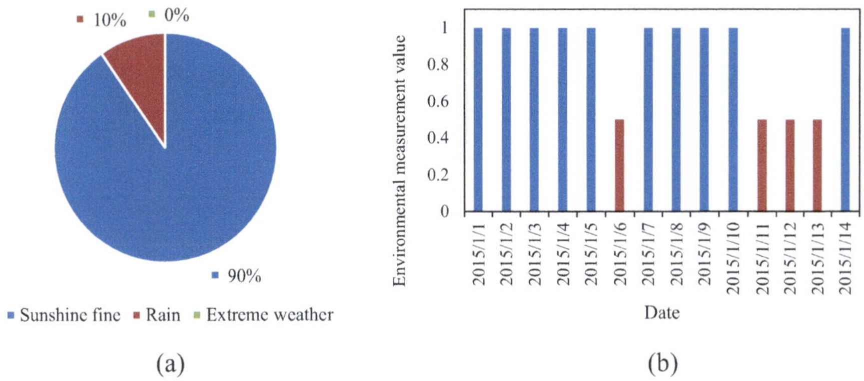

Fig. 2.26 Calculation diagram of the environmental measurement matrix. (**a**) Weather statistics. (**b**) Measurements corresponding to weather

Table 2.14 Structure of the PKET-GCN

Parameter	PKET-GCN
Input dimension	[Batch_size, Number of nodes, 7]
Output dimension	[Batch_size, Number of nodes, 1]
Dynamic node feature dimension	[2, Number of nodes, Number of nodes]
Static node feature dimension	[1, Number of nodes, Number of nodes]
Dilated convolution dimension	[1, 3]
Dilation rate	2
Dynamic convolution dimension	[1, 2, 5], [1, 3]
Activation function	Sigmoid, ReLU, Tanh

- STGCN: The STGCN model uses GCN to extract spatial features and 1D-CNN to extract temporal features.
- STSGCN: The STSGCN model designs a spatial-temporal synchronization modeling mechanism to capture complex local spatial-temporal correlations.
- ETGCN: The ETGCN model proposes a similarity-based fusion method and graph convolution network to fuse multiple graph structures to learn spatial-temporal correlation.

To compare the results fairly, certain model parameters of the six models were unified. The specific model structure settings are shown in Table 2.15.

2.4.6.3 Evaluating Indicator

Four evaluation indexes, including root mean square error (RMSE), root mean square percentage error (RMSPE), mean absolute error (MAE), and mean absolute percentage error (MAPE), were used to evaluate the prediction performance [9]. The formulas of the three evaluation indexes are as follows:

Table 2.15 Parameter settings

Parameter	LSTM	CNN, DCRNN, STGCN, STSGCN, ETGCN
Input dimension	[1,7, Number of nodes]	[7, Number of nodes]
Output dimension	[1, Number of nodes]	[1, Number of nodes]
Training step	200	200
Batch size	30	30
Learning rate	0.0001	0.0001
Time interval	PeMS: 5 min, SZ-taxi: 15 min	PeMS: 5 min, SZ-taxi: 15 min

$$\mathrm{RMSE} = \sqrt{\frac{1}{p}\sum_{i=1}^{p}\left(\boldsymbol{Y}_{\mathrm{Real}}(i) - \boldsymbol{Y}_{\mathrm{Pred}}(i)\right)^2} \tag{2.45}$$

$$\mathrm{RMSPE} = 100\%\sqrt{\frac{1}{p}\sum_{i=1}^{p}\left(\frac{\boldsymbol{Y}_{\mathrm{Real}}(i) - \boldsymbol{Y}_{\mathrm{Pred}}(i)}{\boldsymbol{Y}_{\mathrm{Real}}(i)}\right)^2} \tag{2.46}$$

$$\mathrm{MAE} = \frac{1}{p}\sum_{i=1}^{p}\#\boldsymbol{Y}_{\mathrm{Real}}(i) - \boldsymbol{Y}_{\mathrm{Pred}}(i)\# \tag{2.47}$$

$$\mathrm{MAPE} = \frac{100\%}{p}\sum_{i=1}^{p}\left|\frac{\boldsymbol{Y}_{\mathrm{Real}}(i) - \boldsymbol{Y}_{\mathrm{Pred}}(i)}{\boldsymbol{Y}_{\mathrm{Real}}(i)}\right| \tag{2.48}$$

2.4.6.4 Experimental Performance and Discussion

Prediction Effect of PKET-GCN

The datasets in Sect. 2.4.6 and the parameter settings in Sect. 2.4.6 were used to validate the PKET-GCN model. Figure 2.27 shows the fitting between the predicted value and the actual value of the model in the PeMS test set on a random day. According to the traffic flow data recorded by the sensor, the traffic flow reached its peak at 5 a.m., and the overall traffic flow exhibited a downward trend from 8 p.m. to 2 a.m. By analyzing PeMS04 and PeMS08, the same trend was observed between traffic flow and traffic occupancy. Accordingly, traffic flow decreased along with occupancy of the road. Considering that the sensor selected by PeMS04 was in the trunk line, and that PeMS08 was in the road intersection, the speed data of the former was considered more stable than that of the latter. According to the fitting in Fig. 2.27, PKET-GCN exhibited excellent prediction performance. Figure 2.28 shows the training process of the proposed model in the five traffic flow datasets. With the rise in training time, the prediction error of PKET-GCN in the five datasets gradually converged to a smooth state.

Figure 2.29 shows the location distribution of the three sensors in PeMS03 as well as the corresponding daily average traffic flow prediction. In Fig. 2.29, the

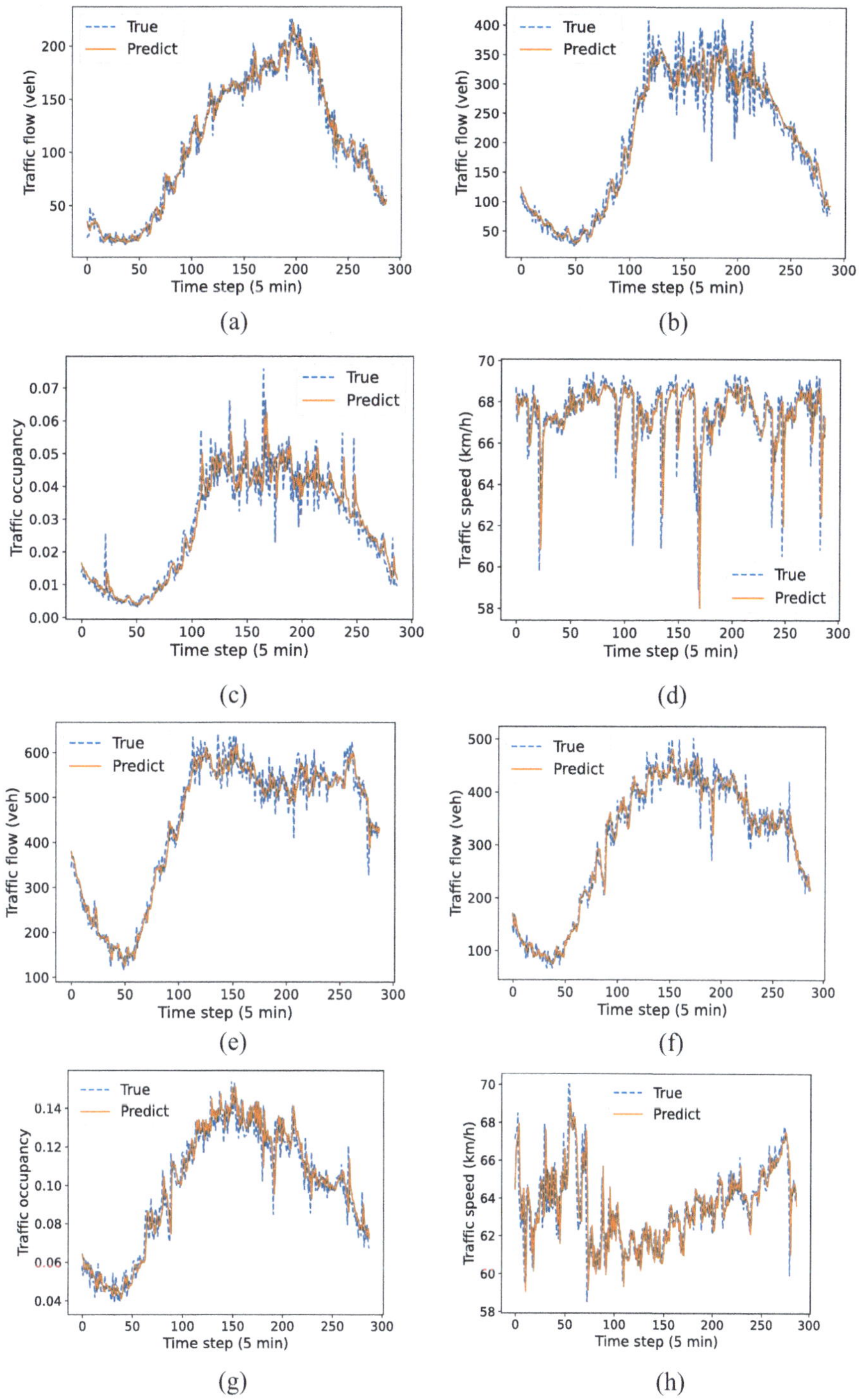

Fig. 2.27 Comparison between predicted and true values of PKET-GCN. (**a**) PeMS03(flow). (**b**) PeMS04(flow). (**c**) PeMS04(occupancy). (**d**) PeMS04(speed). (**e**) PeMS07(flow). (**f**) PeMS08(flow). (**g**) PeMS08 (occupancy). (**h**) PeMS08 (speed)

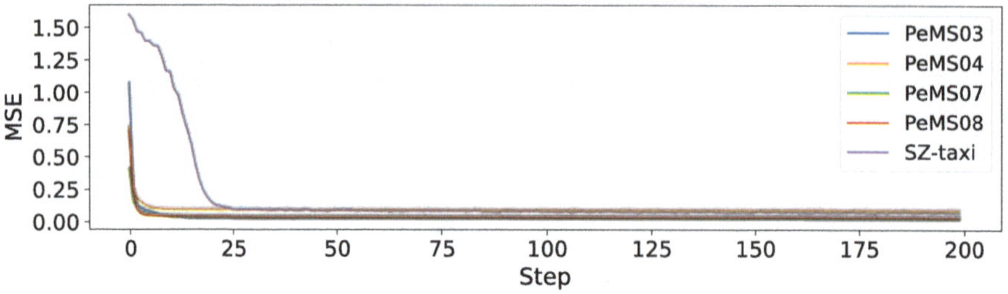

Fig. 2.28 Iteration of the proposed model (times)

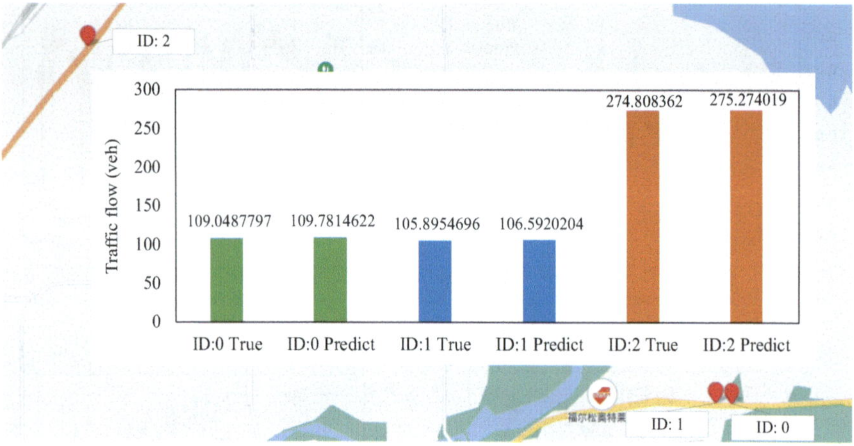

Fig. 2.29 Influence of geographical distribution on traffic flow

average value of 1-day traffic flow of the two sensors with ID = 0 and 1 was equal due to their close distance. The two sensors with ID = 0 and 1 were not on the same trunk line as that of ID = 2, resulting in a significant difference in the traffic flow between the two sensors. In light of the prediction and comparison results, PKET-GCN was shown to fit the traffic flow in different geographical locations and had good prediction performance.

Figure 2.30 shows the impact on traffic flow under different weather conditions. Taking SZ-taxi dataset as an example, Fig. 2.30a shows the change in traffic speed at variable states of weather. According to the results, traffic speed on sunny days was found to be more significant than on rainy days. To demonstrate the ability of the PKET-GCN model in measuring environmental capacity, the traffic speed of two types of weather was predicted. As shown in Fig. 2.30b, the speed model's prediction curve was noted to maintain the same change trend as the actual speed curve, effectively reflecting the change in traffic speed under different weather conditions.

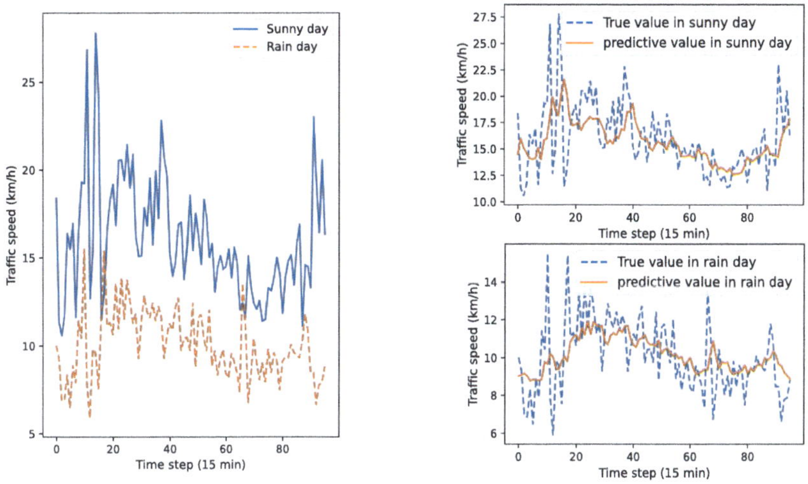

(a) (b)

Fig. 2.30 Influence of weather on traffic flow. (**a**) Traffic flow under different weather conditions. (**b**) Prediction performance of PKET-GCN model under different weather conditions

Table 2.16 The prediction of traffic flow

Classical models							
Datasets	Metrics	LSTM	CNN	DCRNN	STGCN	STSGCN	PKET-GCN
PeMS03	RMSE	35.233	32.987	30.427	30.314	28.873	**27.184**
	RMSPE(%)	31.505	29.010	28.183	27.956	27.523	**26.885**
	MAE	21.132	20.725	18.239	17.485	17.379	**16.387**
	MAPE(%)	21.561	20.773	18.924	17.472	16.882	**16.034**
PeMS04	RMSE	41.732	39.076	38.383	35.423	34.136	**33.741**
	RMSPE(%)	27.762	27.213	26.748	26.213	25.444	**24.216**
	MAE	27.162	27.078	24.926	22.721	21.014	**20.633**
	MAPE(%)	18.598	17.623	17.482	14.688	13.905	**13.200**
PeMS07	RMSE	45.927	40.794	38.856	38.758	38.894	**37.096**
	RMSPE(%)	24.205	23.569	22.847	22.189	21.493	**20.378**
	MAE	30.021	32.141	25.764	25.382	24.768	**23.308**
	MAPE(%)	13.532	12.361	11.693	11.087	10.663	**9.523**
PeMS08	RMSE	34.074	29.528	27.837	27.829	27.555	**26.307**
	RMSPE(%)	22.536	22.083	21.171	20.915	20.706	**20.532**
	MAE	22.373	21.858	17.861	18.035	17.883	**16.895**
	MAPE(%)	14.682	13.790	11.458	11.427	11.709	**9.996**

Comparison of Prediction Performance with Classical Models

Two baseline and three advanced models were chosen as classical models to compare performance with the PKET-GCN model. The PeMS datasets and SZ-taxi dataset were used to train all classical models. Table 2.16 outlines the comparison

Table 2.17 The prediction of traffic speed

Classical models							
Datasets	Metrics	LSTM	CNN	DCRNN	STGCN	STSGCN	PKET-GCN
PeMS04	RMSE	5.011	4.851	4.476	4.254	3.933	**3.741**
	MAE	2.135	1.984	1.904	1.814	1.788	**1.710**
	MAPE(%)	4.960	4.747	4.316	4.108	3.813	**3.542**
PeMS08	RMSE	4.377	4.245	4.102	3.984	3.803	**3.599**
	MAE	1.873	1.824	1.768	1.708	1.640	**1.564**
	MAPE(%)	4.328	4.158	3.991	3.858	3.632	**3.378**
SZ_taxi	RMSE	4.943	4.915	4.893	4.887	4.860	**4.736**
	MAE	3.555	3.541	3.528	3.502	3.491	**3.388**
	MAPE(%)	21.651	20.048	19.570	19.430	19.211	**18.624**

Table 2.18 The prediction of traffic occupancy

Classical models							
Datasets	Metrics	LSTM	CNN	DCRNN	STGCN	STSGCN	PKET-GCN
PeMS04	RMSE	0.0473	0.0471	0.0468	0.0467	0.0466	**0.0465**
	MAE	0.0273	0.0271	0.0267	0.0266	0.0265	**0.0264**
PeMS08	RMSE	0.0438	0.0437	0.0435	0.0434	0.0433	**0.0431**
	MAE	0.0278	0.0276	0.0274	0.0274	0.0273	**0.0272**

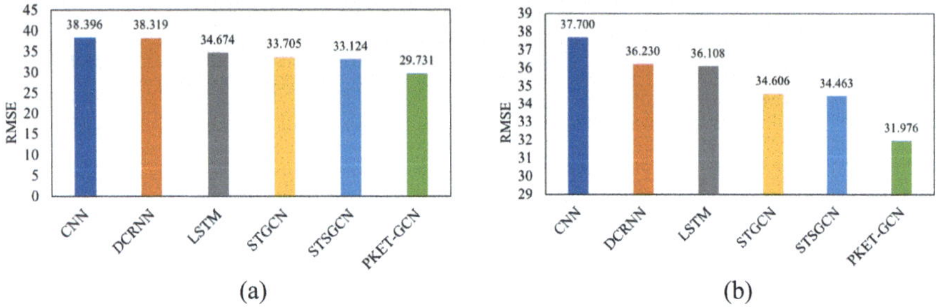

Fig. 2.31 Traffic flow prediction at different time periods. (**a**) 8–10 a.m. (**b**) 10–12 a.m.

results of the flow data of the PeMS datasets. Table 2.17 shows the comparison results of the speed data of the PeMS datasets and SZ-taxi dataset, while Table 2.18 depicts the comparison results of the occupancy data of the PeMS datasets. The corresponding findings show that the proposed PKET-GCN maintained satisfactory performance in all classical models. Compared with other classical models, with MAPE as the evaluation index, the accuracy of PKET-GCN was found to be 8.854% higher than that of LSTM, CNN, DCRNN, STGCN, and STSGCN.

Two time periods (8:00–10:00 a.m. and 10:00–12:00 a.m.) were selected in order to evaluate the peak prediction performance of all comparable models. For example, in the PeMS07 dataset, Fig. 2.31 shows the prediction performance of different models in the corresponding two time periods. Compared with other models, PKET-GCN had better prediction performance in these two time periods.

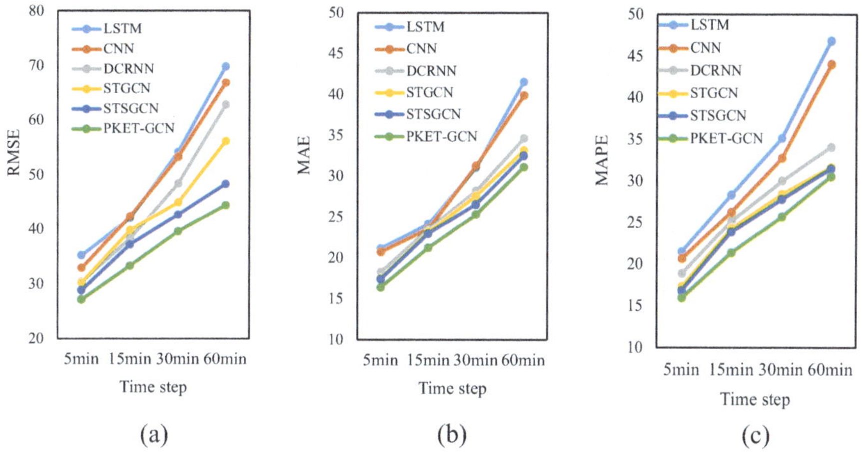

Fig. 2.32 Comparison of prediction performance under multi-step prediction. (**a**) RMSE. (**b**) MAE. (**c**) MAPE

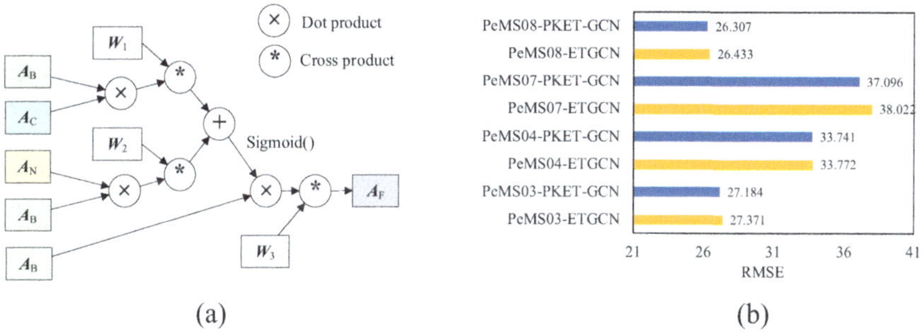

Fig. 2.33 Comparison with the model of improved adjacency matrix. (**a**) Node correlation process of ETGCN. (**b**) Comparison of prediction results

To verify the ability of the proposed model in extracting multi-step time features, the multi-step prediction results of all models were compared. Taking the PeMS03 dataset as the verification set, Fig. 2.32 depicts the multi-step prediction results of the proposed model and other comparable models. The time steps were set to 5 min, 15 min, 30 min, and 60 min, respectively. According to Fig. 2.32, with the increase in time step, the prediction performance of all models was shown to gradually decrease. Among them, the proposed model always maintained excellent prediction performance.

Comparison with the Model of Improved Adjacency Matrix

Different from LSTM, CNN, DCRNN, and STGCN, ETGCN proposed three improved matrices and combined the three matrices into $\boldsymbol{A}_F$, as shown in Fig. 2.33a, to replace the classic adjacency matrix [6]. In order to ascertain the advantages of

Table 2.19 Comparison of space and time complexity

Model	Params (M)	FLOPs (M)	RMSE
STSGCN	**1.095**	**418.184**	27.555
PKET-GCN	1.749	439.516	**26.307**

prior knowledge matrix extraction, the improved adjacency matrix scheme of ETGCN was integrated with the backbone of the proposed model, after which its prediction performance was compared, as shown in Fig. 2.33b. Among the PeMS datasets, the proposed improved adjacency matrix demonstrated better prediction performance.

Comparison of Space and Time Complexity

Past studies have commonly employed params and floating point operations (FLOPs) to measure the complexity of deep learning algorithms [10]. Params correspond to the space complexity and represent the size of the memory occupied. Meanwhile, FLOP corresponds to the time complexity, which can be used to measure the computational complexity and determine the computational time of the algorithm. The smaller the FLOPs, the smaller the computational complexity required. The traffic flow dataset of PeMS08 was used as the test set. The results of PKET-GCN and STSGCN were then compared, as shown in Table 2.19. Compared with STSGCN, the RMSE of PKET-GCN was noted to be 4.529% higher; however, spatial complexity and time complexity increased by 5.973% and 5.101%, respectively.

2.4.6.5 Ablation Analysis

To verify the function of the prior knowledge extraction module on PKET-GCN, the prior knowledge matrix was extracted by dynamically weighting the dynamic node characteristic matrix and the static node characteristic matrix. The dynamic weighting method was defined as: $(\boldsymbol{W}_s\boldsymbol{A}_s) \parallel (\boldsymbol{W}_{d1}\boldsymbol{A}_{d1}) \parallel (\boldsymbol{W}_{d2}\boldsymbol{A}_{d2})$. The model was named PKET-GCN-1, where $\boldsymbol{W}_s$, $\boldsymbol{W}_{d1,}$ and $\boldsymbol{W}_{d2}$ are the dynamic weights corresponding to the three node features matrices, respectively. Taking PeMS03 dataset as the test object, the comparison results are shown in Table 2.20. Compared with the dynamic weight method, the prior knowledge extraction module increased the training time by 1.977%, but improved the prediction accuracy by 1.912%.

The PKET-GCN-ND model was constructed without the dynamic time feature extraction module, while the PKET-GCN-NL model was established without the long-term time feature extraction module. Moreover, the PKET-GCN-NDL model was created without the dynamic and long-term time feature extraction module, while the PKET-GCN-NSA model was made without the self-attention module in order to verify the impact of the time-varying feature extraction module. PeMS03

Table 2.20 Comparison of different improved adjacency matrix methods

Model	Train cost of each epoch(s), batch size = 30	RMSE
PKET-GCN-1	**118.793**	27.714
PKET-GCN	121.142	**27.184**

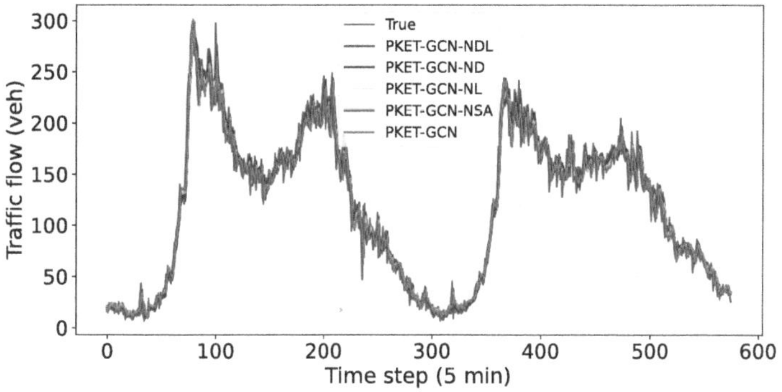

Fig. 2.34 Comparison of prediction results of different components in PeMS03 test set

Table 2.21 Comparison of different time feature extraction components

Model	Train cost of each epoch(s), batch size = 30	RMSE
PKET-GCN-NDL	**117.902**	29.145
PKET-GCN-ND	118.720	28.266
PKET-GCN-NL	118.754	27.762
PKET-GCN-NSA	119.538	27.532
PKET-GCN	121.142	**27.184**

was used as the test dataset, whose comparison results are shown in Fig. 2.34 and Table 2.21. As evident in Fig. 2.34, different variants of PKET-GCN were found to have excellent fitting ability to the actual values in the PeMS03 dataset. In Table 2.21, PKET-GCN was shown to consider both dynamic and long-term temporal features, required the highest training time per round and maintained the highest training accuracy. However, PKET-GCN-NDL training, which does not consider temporal features, was found to have the lowest accuracy. The introduction of dynamic and long-term time features improved the accuracy of PKET-GCN by 6.728%, where the training time of PKET-GCN increased by 2.748%. The introduction of an attention mechanism improved the precision of the proposed model by 1.264%.

The selection of graph convolution layers in the dynamic spatial feature extraction component is also important for model accuracy. Taking PeMS datasets of traffic flow as example, the number of layers of the map volume was set from 1 to 5, in which the RMSE changes are shown in Fig. 2.35. Here, the convolution selection of different datasets was found to be different. If the number of convolution layers was set to a value too small, it caused the model to learn spatial change trends in traffic

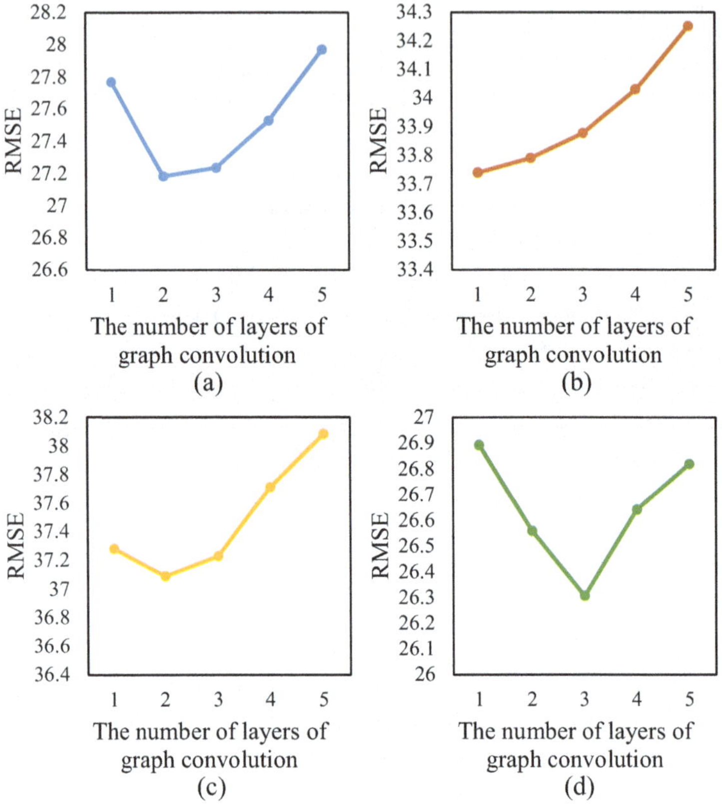

Fig. 2.35 Influence of the number of layers of graph convolution on prediction accuracy. (**a**) PeMS03. (**b**) PeMS04. (**c**) PeMS07. (**d**) PeMS08

flow. If the number of convolution layers was set to a value too large, it led to excessive learning, resulting in reduced accuracy.

2.4.6.6 Discussion

Discussion of Experimental Results

PeMS datasets were used to verify the prediction of PKET-GCN in Sect. 2.4.6. Figure 2.27 shows that traffic flow exhibited a nonlinear fluctuation trend, while the proposed model fit the periodic trend of the traffic flow well. The effective distance between nodes was found to play a role in traffic flow. Figure 2.28 shows that the proposed model demonstrated good convergence in multiple datasets. Meanwhile, Fig. 2.29 shows the relationship between the positions of the three sensors and the traffic flow. According to the prediction and comparison results, the proposed PKET-GCN model can well fit the traffic flow in different geographical locations.

The impact of different weather on traffic flow was also considered in this study. In Fig. 2.30, the traffic flow on sunny days was noted to be significantly greater than that on rainy days. Accordingly, it can be inferred that people are less willing to travel on rainy days than on sunny days. Thus, the proposed model fits the traffic flow in different states of weather very well.

In Sect. 2.4.6, PKET-GCN was compared with existing methods. The comparison results showed that the proposed PKET-GCN maintained adequate performance in all classical models. Compared with other classical models, with MAPE as the evaluation index, the accuracy of the proposed model in traffic prediction was found to be 8.854% higher than that of LSTM, CNN, DCRNN, STGCN, and STSGCN. In previous studies, LSTM has been shown to play an important role in time series prediction tasks and maintained good prediction accuracy. CNN has also demonstrated good recognition accuracy in processing image recognition tasks and performed well in processing Euclidean data. However, the PeMS datasets and SZ-taxi dataset are composed of multiple nodes. The traffic flow between nodes has significant temporal features as well as complex spatial features. In view of this, only one of these features alone cannot improve the prediction accuracy of the models. As road networks in the real world are of non-Euclidean structure, CNN may be prone to losing critical spatial features of correlation when processing such complex data. Unlike LSTM and CNN, DCRNN, STGCN, and STSGCN have better predictability in processing real-world road network data. These three methods consider spatial-temporal features and use the classical adjacency matrix or the linear distance between nodes to measure the correlation between nodes, ignoring the complex dynamic and static correlation features between nodes, which may easily result in the loss of important dynamic spatial-temporal features between nodes. Figures 2.31 and 2.32 both effectively depict that the proposed model performed better than the comparison model in predicting the peak value of traffic flow and multi-step time prediction.

Unlike LSTM, CNN, DCRNN, STGCN, and STSGCN, the ETGCN model adopts a multi adjacency matrix fusion method to comprehensively consider the complex correlation features between nodes. Although this method ignores the impact of the environment outside the node, the fusion method uses a weighted technique, which lacks feature coding and decoding and easily ignores key features. In this study, to substantiate the advantages of prior knowledge matrix extraction, the improved adjacency matrix scheme of ETGCN was integrated with the backbone of the proposed model, and its prediction performance was then compared, as shown in Fig. 2.33b. In the PeMS datasets, the proposed improved adjacency matrix method was shown to have better prediction performance. Compared with the best prediction models of LSTM, CNN, DCRNN, STGCN, STSGCN, and ETGCN, the prediction performance of the proposed model improved by 1.661% on average, demonstrating reasonable practicability and stability. Params and FLOPs were also used to evaluate space-time complexity. Compared with STSGCN, the RMSE of PKET-GCN was found to be 4.529% higher, while spatial complexity and time complexity increased by 5.973% and 5.101%, respectively.

In Sect. 2.4.6, ablation experiments were conduct on the proposed model. To confirm the effectiveness of the prior knowledge extraction module, a dynamic weight method was adopted to fuse the dynamic node and static node features. Compared with the dynamic weight method, the prior knowledge extraction module increased training time by 1.977%, while improving prediction accuracy by 1.912%. In addition, the importance of temporal features in the model was also verified. As seen in Table 2.21, PKET-GCN considered both dynamic and long-term temporal features, required the highest training time per round and maintained the highest training accuracy. The introduction of dynamic and long-term time features improved the accuracy of the proposed model by 6.728%, while the proposed model's training time increased by 2.748%. In conclusion, the proposed model exhibits excellent prediction accuracy in a variety of tasks and shows good practicability.

Limitations

In the ablation experiment detailed in Sect. 2.4.6, the PKET-GCN model's superiority is further shown in traffic flow prediction. However, the PKET-GCN model still has some shortcomings in actual traffic flow prediction. Specifically, Tables 2.17 and 2.18 fully demonstrate that the more complex the structure of the model is, the more training time will be required. However, the prior knowledge matrix effectively considers the features of dynamic and static nodes. In static node features, only the geographical location of the node is considered; in the dynamic node, however, only data correlation and the external weather of the node are considered. In the real world, nodes are subject to additional disturbances. Thus, more external disturbances should be considered in the future to truly reflect disturbances subjected to traffic flow.

2.5 Residual Attention-Enhanced Time-Varying Multi-Factor Graph Convolutional Network (RTM-GCN)

2.5.1 Overview of RTM-GCN

In this paper, a novel RTM-GCN model is proposed to address this issue. As shown in Fig. 2.36, the RTM-GCN model consists of multi-factor graph convolution module, spatial enhancement module, and time-varying module. The multi-factor graph convolution module is used to extract the spatial features of internal and external factors; the spatial enhancement module is used to output global spatial features to enhance the multi-factor graph convolution; the time-varying module is used to enhance the long-term temporal feature and avoid losing historical information. The final predicted output of the model is obtained from the time-varying module. By

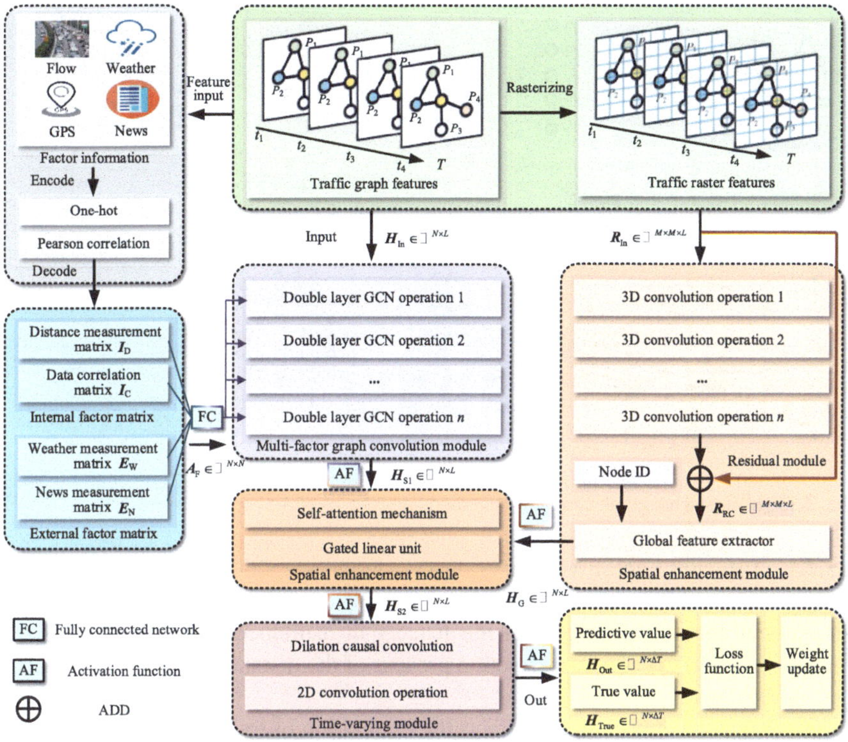

Fig. 2.36 Network structure of RTM-GCN

fusing global spatial features of traffic raster network, RTM-GCN enhances traffic graph network which plays an important role in improving extraction of global features for traffic flow.

2.5.2 *Multi-Factor Graph Convolution Module*

An internal and external factor matrix is established in this paper to better measure the interference of multiple factors on traffic flow. The internal factor matrix, which includes distance measurement matrix and data correlation matrix, is generated based on the latitude and longitude of nodes and historical data. The external factor matrix, which includes weather measurement matrix and news measurement matrix, is generated based on the weather status and news at the location of road network. In this study, a fully connected network is used to learn the contribution of internal factor matrix and external factor matrix to replace classical adjacency matrix. Additionally, a two-layer GCN is utilized to obtain local spatial features of traffic graph feature. The process of multi-factor graph convolution module is shown in Fig. 2.37.

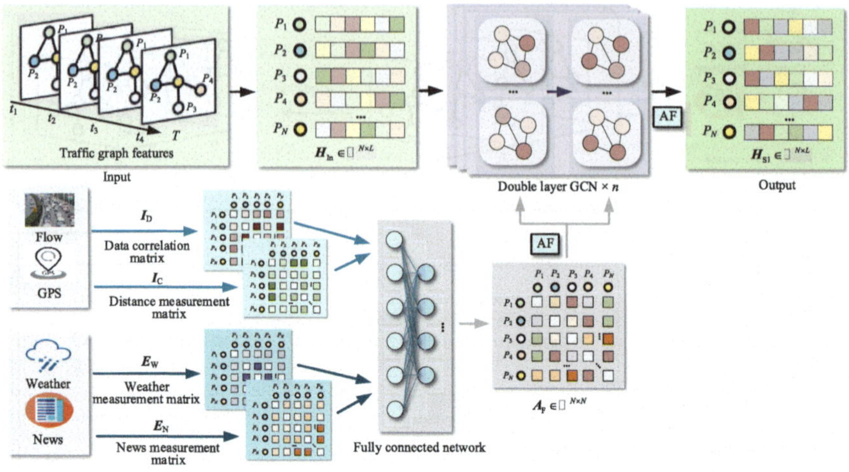

Fig. 2.37 Multi-factor graph convolutional module

2.5.2.1 Internal Factor Matrix

The internal factor matrix includes the distance measurement matrix $\boldsymbol{I}_{\mathrm{D}}$ and the data correlation matrix $\boldsymbol{I}_{\mathrm{C}}$. The construction process of the distance measurement matrix $\boldsymbol{I}_{\mathrm{D}}$ is as follows.

The distance between nodes affects the correlation of traffic flow. Generally, the farther apart nodes are, the smaller their correlation; conversely, the closer they are, the greater their correlation. Based on latitude and longitude, we calculate the physical distance between node Sensor(i) and node Sensor(j), taking into account Earth's curved surface using Haversine formula defined as:

$$\begin{aligned} H_{\text{Hav}}(\theta) = H_{\text{Hav}}\left(\text{Lat}(i) - \text{Lat}(j)\right) \\ + \cos\left(\text{Lat}(i)\right)\cos\left(\text{Lat}(j)\right) H_{\text{Hav}}\left(\left|\text{Lon}(i) - \text{Lon}(j)\right|\right) \end{aligned} \quad (2.49)$$

where $\theta = d/R$, d is the distance between Sensor(i) and Sensor(j), R is the Earth radius, set as 6371 km; Lat(i) and Lat(j) represent the longitude of Sensor(i) and Sensor(j), and Lon(i) and Lon(j) represent the latitude of Sensor(i) and Sensor(j), respectively; $H_{\text{Hav}}(\theta) = (1 - \cos(\theta))/2$. Based on this equation, the physical distance between all nodes can be obtained. A distance threshold α is established. If d_{ij}, the distance between Sensor(i) and Sensor(j), is greater than α, then $\boldsymbol{I}_{\mathrm{D}}(i, j) = 0$; otherwise, it equals 1. The distance measurement discrimination equation can be obtained as follows:

$$D_I(i,j) = \begin{cases} 0 & d_{ij} > \alpha \\ 1 & d_{ij} \le \alpha \end{cases} \quad (2.50)$$

According to the distance measurement discrimination formula mentioned above, a distance measurement matrix $\boldsymbol{I}_{\mathrm{D}} \in R^{N \times N}$ can be generated. The construction process of the data correlation matrix $\boldsymbol{I}_{\mathrm{C}}$ is as follows.

There exists a correlation between the data of nodes. If the traffic flow information change trend of Sensor(i) and Sensor(j) are similar, it represents that the traffic flow information of these two nodes are highly correlated. Therefore, A data correlation matrix is established in this paper to reflect the correlation between nodes. The Pearson correlation coefficient is widely used to measure the connectivity of two time series data, defined as follows:

$$\mathrm{Pea}(i,j)=\frac{\sum_{t=1}^{T_{\mathrm{H}}}\left(h_{i,t}-\mathrm{Mean}\left(\boldsymbol{h}_i\right)\right)\left(h_{j,t}-\mathrm{Mean}\left(\boldsymbol{h}_j\right)\right)}{\mathrm{STD}\left(\boldsymbol{h}_i\right)\mathrm{STD}\left(\boldsymbol{h}_j\right)} \tag{2.51}$$

where Mean(•) and STD(•) are the operations of calculating the mean and standard deviation, respectively. $h_{i,t}$ represents the traffic flow information of Sensor(i) at time t. Pea(i, j) outputs a correlation measurement value between two nodes, with a range of (−1, 1). The closer the numerical value is to 1 or −1, the greater the positive or negative correlation at that time. Based on this equation, this paper obtains a correlation measurement for all nodes and generate a data correlation matrix $\boldsymbol{I}_{\mathrm{C}} \in R^{N \times N}$.

2.5.2.2 External Factor Matrix

The external factor matrix includes the weather measurement matrix $\boldsymbol{E}_{\mathrm{W}}$ and news measurement matrix $\boldsymbol{E}_{\mathrm{N}}$. Weather data and news data are crucial to traffic flow information. Severe weather can reduce people's willingness to travel, while city marathon news or regional road maintenance can remind people to change routes or not travel. Both of these situations can cause non-periodic changes in traffic flow. The influence of weather data and news data on the traffic flow information of Sensor(i) and Sensor(j) is an external interference factor that belongs to text data. It is essential to quantify them numerically.

The impact of weather and news data on travel is encoded using the one-hot method in this paper. Taking weather data as an example, the willingness to travel between Sensor(i) and Sensor(j) is encoded into a vector of length 3 based on weather conditions, such as [*comfortable travel, normal travel, difficult travel*]. If the weather makes it difficult for people to travel between Sensor(i) and Sensor(j), then the encoding feature $\boldsymbol{En}_{\mathrm{We}}(i,j)$ at this time is [0, 0, 1]. If the decoding matrix of weather is defined as $\boldsymbol{W}_{\mathrm{We}}$ = [0.9, 0.5, 0.1], then the specific weather measurement value between Sensor(i) and Sensor(j) can be obtained from $\boldsymbol{En}_{\mathrm{We}}\boldsymbol{W}_{\mathrm{We}}^{\mathrm{Tr}}$ as 0.1 by using Tr for transpose operation.

According to the above method, this paper can generate a weather measurement matrix $\boldsymbol{E}_{\mathrm{W}} \in R^{N \times N}$ between all nodes. The acquisition of the news measurement matrix $\boldsymbol{E}_{\mathrm{N}}$ is obtained in the same way as the weather measurement matrix. By defining corresponding encoding features $\boldsymbol{En}_{\mathrm{N}}$ and decoding matrix $\boldsymbol{W}_{\mathrm{N}}$, it is possible to output a news measurement matrix $\boldsymbol{E}_{\mathrm{N}} \in R^{N \times N}$ between all nodes.

2.5.2.3 Multi-Factor Feature Fusion Extraction

Based on the difficulty of effectively measuring the dynamic relationship features between nodes using the classical adjacency matrix, this paper proposes a multi-factor fusion feature $\boldsymbol{A}_{\mathrm{F}}$ by fusing the characteristics of internal and external factor matrices based on a fully connected network. Unlike the adjacency matrix, $\boldsymbol{A}_{\mathrm{F}}$ can effectively measure the influence of multiple factors to better reflect the interrelationships between nodes. The specific operation is as follows:

$$\boldsymbol{A}_{\mathrm{F}} = f_{\mathrm{AF}}\left(\mathrm{FC}_1\left(\boldsymbol{I}_{\mathrm{D}}\right)+\mathrm{FC}_2\left(\boldsymbol{I}_{\mathrm{C}}\right)+\mathrm{FC}_3\left(\boldsymbol{E}_{\mathrm{W}}\right)+\mathrm{FC}_4\left(\boldsymbol{E}_{\mathrm{N}}\right)\right) \tag{2.52}$$

where FC (•) represents fully connected operation and f_{AF} represents activation function. Considering the excellent performance of graph convolutional network in traffic graph network, this paper uses a two-layer GCN to preliminarily extract spatial features of $\boldsymbol{H}_{\mathrm{In}} \in \mathrm{R}^{N \times L}$. The specific operations are as follows:

$$\boldsymbol{X}_{2\mathrm{GCN}}^{l+1} = f_{\mathrm{AF}}\left(\boldsymbol{A}_{\mathrm{F}}\left(\boldsymbol{A}_{\mathrm{F}}\boldsymbol{X}_{2\mathrm{GCN}}^{l}\boldsymbol{W}_{2\mathrm{GCN}}^{1}\right)\boldsymbol{W}_{2\mathrm{GCN}}^{2}\right) \tag{2.53}$$

where $\boldsymbol{X}_{2\mathrm{GCN}}^{l+1}$ and $\boldsymbol{X}_{2\mathrm{GCN}}^{l}$ represent the input and output of the double-layer graph convolution operation in the l-th layer, $\boldsymbol{W}_{2\mathrm{GCN}}^{1}$ and $\boldsymbol{W}_{2\mathrm{GCN}}^{2}$ are the weights of the first and second layers of double-layer graph convolution. Finally, we obtain the output $\boldsymbol{H}_{\mathrm{S1}} \in \mathrm{R}^{N \times L}$ for multi-factor spatial features.

2.5.3 *Spatial Enhancement Module*

Enhancing the global spatial feature extraction ability of graph convolution is crucial for increasing the performance of traffic flow prediction. As shown in Fig. 2.38, this paper proposes a novel spatial enhancement module to enhance the global spatial features of graph convolution. The spatial enhancement module consists of residual convolutional networks and self-attention gated linear units.

Given the excellent global spatial feature extraction ability of CNN in traffic raster networks, this paper constructs a residual convolutional network to extract the global spatial features. The residual convolutional network includes 3D convolution operations and residual operations, where 3D convolution operations are defined as:

$$\boldsymbol{X}_{3\mathrm{DConv}}^{l+1} = f_{\mathrm{AF}}\left(\boldsymbol{W}_{3\mathrm{DConv}}^{l} * \boldsymbol{X}_{3\mathrm{DConv}}^{l}\right) \tag{2.54}$$

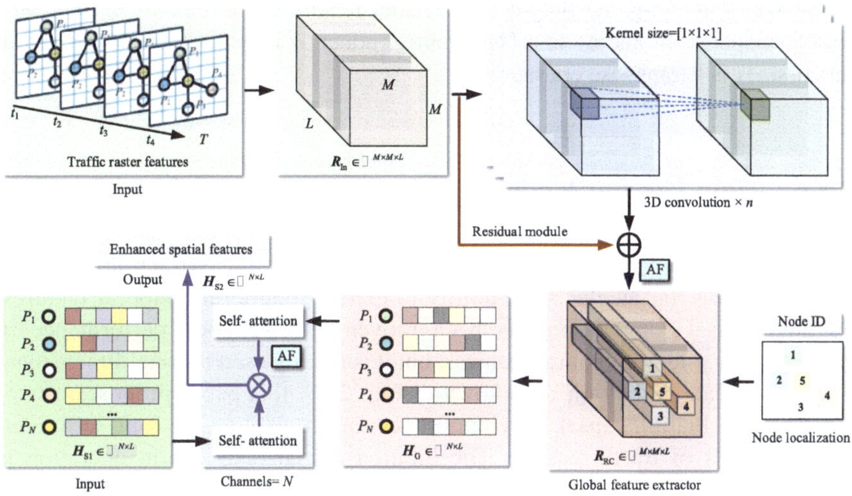

Fig. 2.38 Spatial enhancement module

where $\boldsymbol{X}_{\mathrm{3DConv}}^{l}$ and $\boldsymbol{X}_{\mathrm{3DConv}}^{l+1}$ are the input and output of the l-th layer convolution operation, respectively, and $\boldsymbol{W}_{\mathrm{3DConv}}^{l}$ is the weight of the l-th layer convolution operation. In order to avoid gradient vanishing and capture the impact of local changes on data, a residual operation is introduced to update the above operations as follows:

$$\boldsymbol{X}_{\mathrm{3DConv}}^{l+1} = \boldsymbol{X}_{\mathrm{3DConv}}^{l} + f_{\mathrm{AF}}\left(\boldsymbol{W}_{\mathrm{3DConv}}^{l} {}^{*} \boldsymbol{X}_{\mathrm{3DConv}}^{l}\right) \tag{2.55}$$

the final output of the residual convolutional network is $\boldsymbol{R}_{\mathrm{RC}} \in \mathrm{R}^{M \times M \times L}$.

To apply the global spatial feature $\boldsymbol{R}_{\mathrm{RC}}$ output by residual convolutional network to each node, a global spatial feature extractor is constructed. According to the node number, the raster data $\boldsymbol{R}_{\mathrm{RC}}$ is traversed and the global spatial feature $\boldsymbol{H}_{\mathrm{G}} \in \mathrm{R}^{N \times L}$ is obtained based on the node order of traffic feature matrix $\boldsymbol{H}$.

The self-attention gated linear unit is designed to better integrate the contributions of multi-factor spatial feature $\boldsymbol{H}_{\mathrm{S1}}$ and global spatial feature $\boldsymbol{H}_{\mathrm{G}}$, and the operations are defined as:

$$\boldsymbol{H}_{\mathrm{S2}} = f_{\mathrm{Att}}\left(\boldsymbol{H}_{\mathrm{S1}}\right) \otimes \sigma\left(f_{\mathrm{Att}}\left(\boldsymbol{H}_{\mathrm{G}}\right)\right) \tag{2.56}$$

$$f_{\mathrm{Att}}\left(\boldsymbol{X}\right) = \left(\boldsymbol{X}\right)^{\mathrm{Tr}} \cdot \mathrm{Att}\left(\boldsymbol{X}\right) \tag{2.57}$$

$$\mathrm{Att}\left(\boldsymbol{X}\right) = \frac{\mathrm{e}^{\left(\boldsymbol{X} \cdot \left(\boldsymbol{X}\right)^{\mathrm{Tr}}\right)}}{\sum_{1}^{T} \mathrm{e}^{\left(\boldsymbol{X} \cdot \left(\boldsymbol{X}\right)^{\mathrm{Tr}}\right)}} \tag{2.58}$$

where $H_{S2} \in R^{N \times L}$ is the output after spatial feature enhancement, f_{Att} is a self-attention operation, and ⊗ is a Hadamard product, σ It is the sigmoid activation function, Tr is a transpose operation.

2.5.4 *Time-Varying Module*

The spatial enhancement module effectively improves the extraction of global spatial features, but neglects the ability to extract long-term temporal features. Dilation causal convolution has been proven to have good performance in extracting long-term temporal features, but it is mainly used for one-dimensional time series data extraction. As shown in Fig. 2.39, this paper proposes a two-dimensional dilation causal convolution network to explore the long-term temporal features of traffic graphs. The dilation causal convolution operation is defined as:

$$X_{\mathrm{TCN}}^{l+1} = f_{\mathrm{AF}}\left(W_{\mathrm{TCN}}^{l} X_{\mathrm{TCN}}^{l} + b_{\mathrm{TCN}}^{l}\right) \tag{2.59}$$

where X_{TCN}^{l+1} and X_{TCN}^{l} are the output and input of dilated causal convolution in the l-th layer, respectively. W_{TCN}^{l} and b_{TCN}^{l} are the weight and bias term of dilated causal convolution respectively. Given input $X_{\mathrm{TCN}}^{l} \in \mathbb{R}^{N \times L}$, where dilation rate is set to $1 \times \gamma$, through multiple convolution operations, the final predicted output is $Y_P \in R^{N \times K}$, where K is the length of sequence to be predicted. Mean Squared Error (MSE) is used as a Loss function to train RTM-GCN:

$$\mathrm{MSE} = \frac{\sum_{i=1}^{p} \left|Y_{\mathrm{R}}(i) - Y_{\mathrm{P}}(i)\right|^{2}}{p} \tag{2.60}$$

where Y_R and Y_P is the true value and predicted value of traffic flow, and p is the number of traffic flow training samples. The specific training process of the model is shown in Algorithm 2.

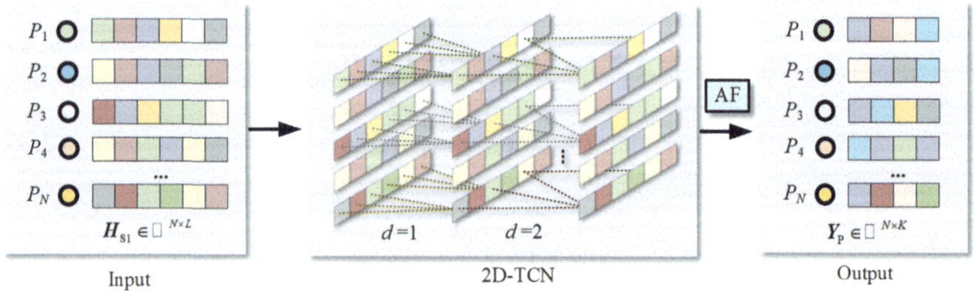

Fig. 2.39 Time-varying module

Algorithm 2 RTM-GCN Training Algorithm

Input: Number of nodes: N; the length of the input sequence: L; the historical traffic graph features: $\boldsymbol{H}_{\text{In}} \in \mathrm{R}^{N \times L}$; the length and width of the raster area: M; the historical traffic raster sequence: $\boldsymbol{R}_{\text{In}} \in \mathrm{R}^{M \times M \times L}$; time interval: ΔT

Output: Learned RTM-GCN model

// Initialize parameters
MFGCM (θ_1) // weight initialization of multi-factor graph convolution module
SEM (θ_2) // weight initialization of spatial enhancement module
TVM (θ_3) // weight initialization of time-varying module
Generation of internal factor matrix $\boldsymbol{I}_{\text{D}} \in \mathrm{R}^{N \times N}$ and $\boldsymbol{I}_{\text{C}} \in \mathrm{R}^{N \times N}$
Generation of external factor matrix $\boldsymbol{E}_{\text{W}} \in \mathrm{R}^{N \times N}$ and $\boldsymbol{E}_{\text{N}} \in \mathrm{R}^{N \times N}$
// train the model
for epoch in max epoch **do**
  $\boldsymbol{A}_{\text{F}} \in \mathrm{R}^{N \times N}$ = fully connected ($\boldsymbol{I}_{\text{D}}, \boldsymbol{I}_{\text{C}}, \boldsymbol{E}_{\text{W}}, \boldsymbol{E}_{\text{N}}$)
  local spatial feature output $\boldsymbol{H}_{\text{S1}} \in \mathrm{R}^{N \times L}$ = MFGCM($\boldsymbol{A}_{\text{F}}, \boldsymbol{H}_{\text{In}}$)
  global spatial feature output $\boldsymbol{H}_{\text{S2}} \in \mathrm{R}^{N \times L}$ = SEM($\boldsymbol{H}_{\text{S1}}, \boldsymbol{R}_{\text{In}}$)
  prediction result output $\boldsymbol{H}_{\text{Out}} \in \mathrm{R}^{N \times \Delta T}$ = TVM($\boldsymbol{H}_{\text{S2}}$)
  calculation of loss value MSE ($\boldsymbol{H}_{\text{Out}}, \boldsymbol{H}_{\text{True}}$)
  update weights
end for
output the learned RTM-GCN model

2.5.5 *Experiments*

2.5.5.1 Dataset Description

In this paper, five datasets are used: PeMS03, PeMS04, PeMS07, PeMS08, and SZ-TAXI.

PeMS (03–08): The Performance Measurement System (PeMS) dataset provides a unified traffic data database collected by Caltrans on highways in California, USA. The data collection interval for PeMS (03–08) is 5 min. This paper mainly uses the traffic flow of PeMS (03–08) as test data.

SZ-TAXI: This dataset is collected from taxi data in Shenzhen, China. The recorded time interval is 15 min, and the data represents traffic speed.

Table 2.22 provides specific descriptions of the five datasets. The latitude and longitude of nodes on roads, weather conditions, and news events are crucial to the accuracy of the model. This paper obtained node latitude and longitude data from "https://pems.dot.ca.gov/," weather information from "http://tianqi.2345.com/", and news events (holidays and traffic accidents) from "https://www.google.com/" based on node location information. After obtaining weather and news data, this paper transforms them into external factor matrices similar to adjacency matrices to reflect weather and news metrics between each node. Specifically, this paper sets the measurement weights for both weather and news to [0.9, 0.5, 0.1], and the time

Table 2.22 Dataset description

Datasets	PeMS03	PeMS04	PeMS07	PeMS08	SZ-TAXI
Location	In California, USA	In California, USA	In California, USA	In California, USA	In Shenzhen, China
Date	9/1–11/30/2018	1/1–2/28/2018	5/1–8/31/2017	7/1–8/31/2016	1/1–1/31/2015
Time interval	5 min	5 min	5 min	5 min	15 min
Data type	Flow	Flow	Flow	Flow	Speed
Nodes	358	307	883	170	156
Intervals	26,208	16,992	28,224	17,856	2976

Table 2.23 Numerical conversion of weather and news

Weather	News	Value
Clear, cloudy, drizzle, etc.	Entertainment, holiday, etc.	0.9
Haze, snow, heavy rain, etc.	Traffic jam, road maintenance, etc.	0.5
Fog, hail, tornado, hurricane, etc.	Pandemic, terrorism, etc.	0.1

interval of the external factor matrix is consistent with the interval of traffic in the dataset. The specific values corresponding to weather and news are shown in Table 2.23. After collecting all necessary data, preprocessing was conducted to avoid missing or redundant values before standardization was applied.

2.5.5.2 Experimental Settings

Hyper-Parameter Settings

The deep learning framework PyTorch was used to implement the RTM-GCN and other deep learning models in this paper. The experiment was conducted on a system with an NVIDIA RTX 3080, Intel i7-12700, and Windows 11 operating system. Adam optimizer was used for RTM-GCN. The maximum training iterations for RTM-GCN were set to be 800, with an initial learning rate of 0.0001 and a batch size of 30. The division ratio of the training set and testing set in Sect. 2.5.5 is 7:3. This paper equates the test set with the validation set for performance verification during model training process. Table 2.24 shows other important parameters of RTM-GCN.

Baseline Models

Fourteen comparative models were selected for this study. The details are as follows:

(a) **ARIMA**: Auto-regressive integrated moving average model is a classic time series analysis approach for predicting tasks.

Table 2.24 Structural parameters of RTM-GCN

Parameter	RTM-GCN
Input dimension	[30, Number of nodes, 12]
Output dimension	[30, Number of nodes, 12]
Dilation rate	[1, 2]
2D convolutional kernel	[1 × 2]
3D convolutional kernel	[1 × 1 × 1]
α	2
Number of self-attention channels	Number of nodes
Activation function	Sigmoid, ReLU, Tanh

(b) **LSTM**: By improving the RNN model to effectively reduce the occurrence of gradient vanishing or exploding, LSTM can be used to extract temporal features.
(c) **CNN**: The CNN model uses regular convolution kernel to extract spatial features of Euclidean structure.
(d) **DCRNN**: The DCRNN model uses two-way random walk to model spatial dependence and encoder–decoder architecture to model temporal dependence.
(e) **STGCN**: The STGCN model uses GCN to extract spatial features and 1D-CNN to extract temporal features.
(f) **ASTGCN**: The ASTGCN model uses spatial-temporal attention mechanism to extract multiperiod spatial-temporal feature.
(g) **STSGCN**: The STSGCN model designs a spatial-temporal synchronization modeling mechanism to capture complex local spatial-temporal correlations.
(h) **STFGNN**: The STFGNN model proposes a data-driven method to generate a "temporal graph" to compensate for the correlation that may not be reflected by the spatial graph.
(i) **Auto-DSTSGN**: The Auto-DSTSGN model proposes an Auto DSTSG module to capture short-term and long-term spatial-temporal correlations by stacking deeper layers with expansion factors in an increasing order.
(j) **MVSTT**: The MVSTT model learns spatial-temporal features from three views and combines GNN-based modules with converters for spatiotemporal information fusion.
(k) **MD-GCN**: The MD-GCN model uses gated temporal convolution combined with GraphSAGE and MGCN to extract global and local correlations of traffic flow.
(l) **DST-GraphSAGE**: The DST-GraphSAGE model uses the spatial-temporal GraphSAGE module in combination with an attention mechanism and dilated causal convolution to extract local and long-term spatial-temporal features of traffic flow.
(m) **GMAN**: The GMAN model uses spatial and temporal attention mechanisms to model spatial and temporal correlations, allowing for better long-term prediction.

(n) **TFormer**: The TFormer model is based on a global encoder and a global local decoder to extract global and local spatial features.

ARIMA is a typical model-driven method for time series prediction, while LSTM and CNN are two typical deep learning methods for single feature extraction. DCRNN, STGCN, ASTGCN, STSGCN, STFGNN, Auto-DSTSGN, and MVSTT are the more popular traffic flow prediction models based on GNN. MD-GCN and DST-GraphSAGE are the traffic flow prediction models fused with the GraphSAGE module. GMAN and TFormer are the traffic flow prediction models fused with transformer module. The above five comparative model categories were chosen to better demonstrate the advantages of the RTM-GCN model in each domain.

Evaluation Metrics

This paper uses three commonly used metrics in traffic flow prediction tasks, Root Mean Square Error (RMSE), Mean Absolute Error (MAE), and Mean Absolute Percentage Error (MAPE), to evaluate the prediction accuracy of the proposed model. The equations for the three evaluation metrics are as follows:

$$\mathrm{RMSE} = \sqrt{\frac{1}{p}\sum_{i=1}^{p}\left(\boldsymbol{Y}_{\mathrm{R}}(i) - \boldsymbol{Y}_{\mathrm{P}}(i)\right)^2} \tag{2.61}$$

$$\mathrm{MAE} = \frac{1}{p}\sum_{i=1}^{p}\#\boldsymbol{Y}_{\mathrm{R}}(i) - \boldsymbol{Y}_{\mathrm{P}}(i)\# \tag{2.62}$$

$$\mathrm{MAPE} = \frac{100\%}{p}\sum_{i=1}^{p}\left|\frac{\boldsymbol{Y}_{\mathrm{R}}(i) - \boldsymbol{Y}_{\mathrm{P}}(i)}{\boldsymbol{Y}_{\mathrm{R}}(i)}\right| \tag{2.63}$$

In addition to three metrics for evaluating prediction accuracy, the spatial and temporal complexity of the model is evaluated using Params and FLOPs, respectively. Params and FLOPs are commonly used to measure the spatial occupancy and temporal efficiency of models, with lower values being better.

Construction of Traffic Raster Feature

For publicly available datasets that can collect node latitude and longitude, this paper directly uses raster division to construct traffic raster feature. According to the actual distribution of nodes, the road is divided into equally sized grids, and the sum of traffic flow statistics for each raster represents the traffic status of that area. Figure 2.40a shows the distribution of all nodes in PeMS03 in a raster network based on latitude and longitude data. For publicly available datasets where it is difficult to obtain latitude and longitude data, this paper arranges nodes in order and combines them into a three-dimensional tensor with dimensions $N \times 1 \times T_{\mathrm{in}}$ as input

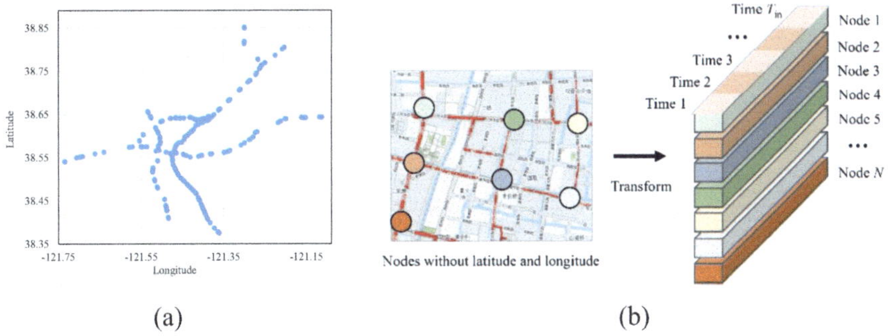

Fig. 2.40 Construction of traffic raster feature. (**a**) Node distribution of PeMS03. (**b**) Construction methods without positioning nodes

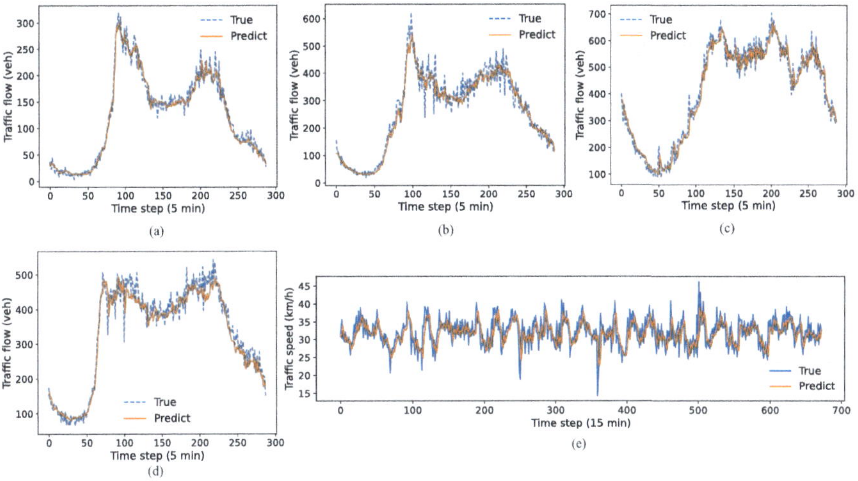

Fig. 2.41 Prediction performance of RTM-GCN. (**a**) PeMS03, RMSE = 12.659. (**b**) PeMS04, RMSE = 34.071. (**c**) PeMS07, RMSE = 37.452. (**d**) PeMS08, RMSE = 30.775. (**e**) SZ-TAXI, RMSE = 3.209

for traffic grid feature extraction, where T_{in} is the length of historical data cycles. Figure 2.40b shows a schematic diagram of constructing raster feature for non-located nodes.

2.5.5.3 Experimental Results

Prediction Performance of RTM-GCN

This paper randomly selects the predicted results of a node in the test set to demonstrate the performance of RTM-GCN. The fitting of RTM-GCN on the PeMS (03–08) and SZ-TAXI datasets is shown in Fig. 2.41. From Fig. 2.41a, d, it can be

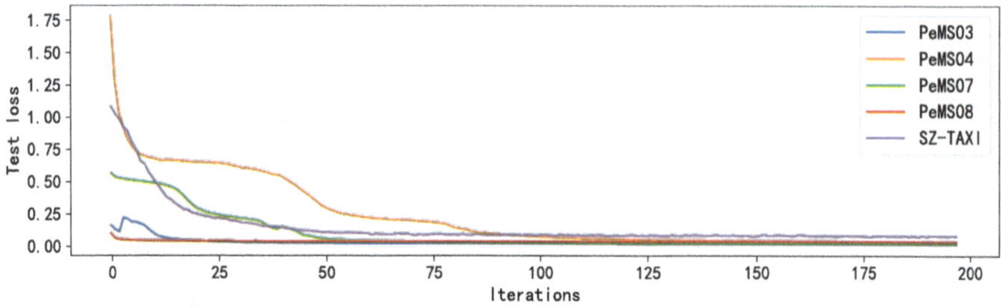

Fig. 2.42 Convergence of RTM-GCN

observed that there are obvious traffic peaks and valleys in the PeMS (03–08) dataset, with traffic volume gradually increasing from 4 am and reaching its first peak at around 8:30 am, then decreasing around 6 pm. Figure 2.41e shows the speed fitting situation of a seven-day test set in SZ-TAXI, which has greater data fluctuations. The fitting situation shown in Fig. 2.41 demonstrates that RTM-GCN proposed in this paper can fit long-term traffic data well and has good predictive performance.

This paper shows the iteration of RTM-GCN within 200 times in five datasets to verify scalability. As shown in Fig. 2.42, the convergence of RTM-GCN is the fastest before the 19th iteration, gradually reaches a smooth state between the 19th and 86th rounds, and is basically smooth after the 86th round.

The impact of news events on traffic flow is enormous. Taking node #313487 in the PeMS03 dataset as an example, on September 3, 2018, which was a Monday, the normal traffic volume should have been high as shown in Fig. 2.43b, but the actual situation was the opposite as shown in Fig. 2.43a. This paper discovered through mining news events that September 3 was Labor Day in the United States and there were also emergency road repairs in the area, resulting in different traffic trends compared to previous days. The RTM-GCN proposed in this paper measures the interference experienced by nodes by mining internal and external factors features of nodes. Figure 2.43c shows RTM-GCN's prediction for node #313487 from November 17 to November 23, 2018. It can be seen that the trend on November 22 and November 23 is different from that of Fig. 2.43b because these 2 days are Thanksgiving Day in America. On November 23, there was also a traffic accident near this node. The model proposed in this paper can fit well with traffic flow under special events and has good application value.

Performance Comparison Between RTM-GCN and Baseline Models

This paper selects PeMS (03–08) as the test data to demonstrate the performance comparison between RTM-GCN and ten baseline models. Table 2.25 shows the average prediction accuracy of RTM-GCN and ten baseline models in the test set of PeMS (03–08). Using RMSE as an evaluation metric, RTM-GCN proposed in this paper improves the accuracy of the best-performing model in the baseline models

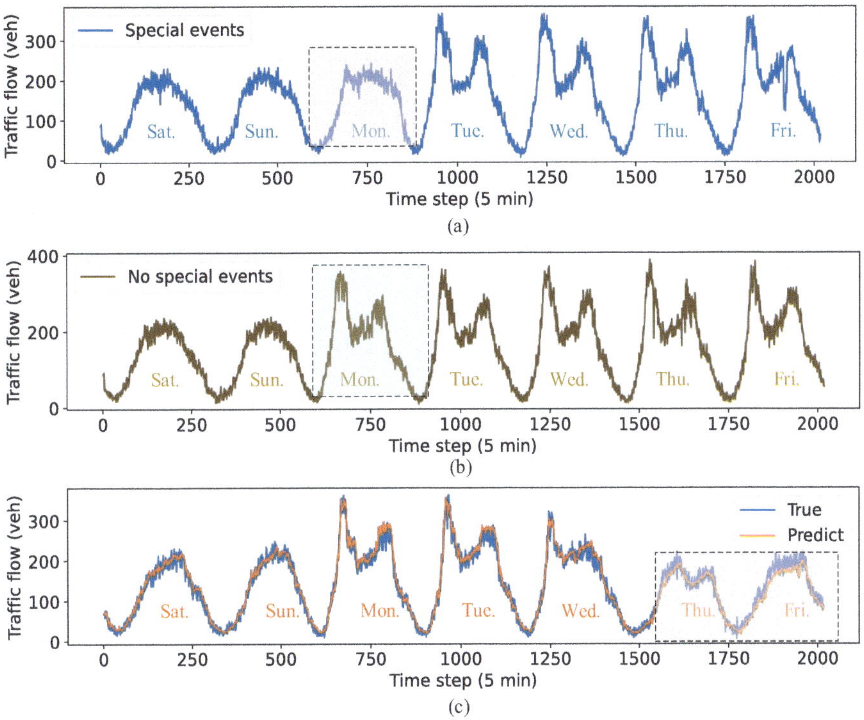

Fig. 2.43 Prediction performance of RTM-GCN under special events. (**a**) Special events (September 3, 2018—Labor day). (**b**) No special events. (**c**) Fitting situation (November 22–23, 2018—Thanksgiving day)

by 6.362%. Table 2.25 shows that RTM-GCN exhibits excellent results in all four datasets.

To test the model's perception ability of traffic flow at the key time point from 10:00 am to 12:00 am, this paper selects SZ-TAXI as the test data to demonstrate the average fitting performance of all models, as shown in Fig. 2.44a. The comparison models include traffic flow prediction models based on GNN, GraphSAGE, and Transformer, RTM-GCN shows better prediction performance during peak traffic periods. To test the model's perception ability for special holidays, this paper selects Thanksgiving period in PeMS03 to demonstrate the average fitting ability of models, as shown in Fig. 2.44b The GNN-based traffic flow prediction models were selected as the test models, and it can be seen from Fig. 2.44 that RTM-GCN is more capable of sensing special events.

Extracting long-term features is crucial as it determines whether a model can maintain high accuracy over longer prediction periods. This paper uses PeMS03 as the test data to demonstrate the predictive performance of all models under multiple prediction periods. The prediction period is divided into 5 min, 15 min, 30 min, and 60 min to better verify the long-term predictive ability of the models. The results are shown in Fig. 2.45. It can be observed that as the length of the prediction period

Table 2.25 Compare with the average results of the baseline models

Classical models												
Datasets	Metrics	ARIMA	LSTM	CNN	DCRNN	STGCN	ASTGCN	STSGCN	STFGNN	MVSTT	Auto-DSTSGN	RTM-GCN
PeMS03	RMSE	41.587	35.233	32.987	30.427	30.314	29.676	28.873	28.342	26.413	25.170	**21.362**
	MAE	26.896	21.132	20.725	18.239	17.485	17.693	17.379	16.773	15.711	**14.592**	15.380
	MAPE(%)	27.314	21.561	20.773	18.924	17.472	19.289	16.882	16.309	15.763	14.226	**13.767**
PeMS04	RMSE	57.139	41.732	39.076	38.383	35.423	35.222	34.136	31.891	31.228	30.488	**28.954**
	MAE	37.311	27.162	27.078	24.926	22.721	22.933	21.014	19.836	19.549	18.854	**18.387**
	MAPE(%)	26.050	18.598	17.623	17.482	14.688	16.577	13.905	13.021	13.113	13.213	**12.980**
PeMS07	RMSE	55.348	45.927	40.794	38.856	38.758	42.571	38.894	35.878	34.467	33.024	**32.483**
	MAE	37.849	30.021	32.141	25.764	25.382	28.352	24.768	22.087	21.071	20.089	**19.225**
	MAPE(%)	17.366	13.532	12.361	11.693	11.087	13.889	10.663	9.253	9.174	8.571	**7.927**
PeMS08	RMSE	41.671	34.074	29.528	27.837	27.829	28.380	27.555	26.231	25.248	23.772	**22.904**
	MAE	28.123	22.373	21.858	17.861	18.035	18.591	17.883	16.642	15.924	14.749	**14.188**
	MAPE(%)	18.259	14.682	13.790	11.458	11.427	13.246	11.709	10.608	10.428	9.451	**9.260**

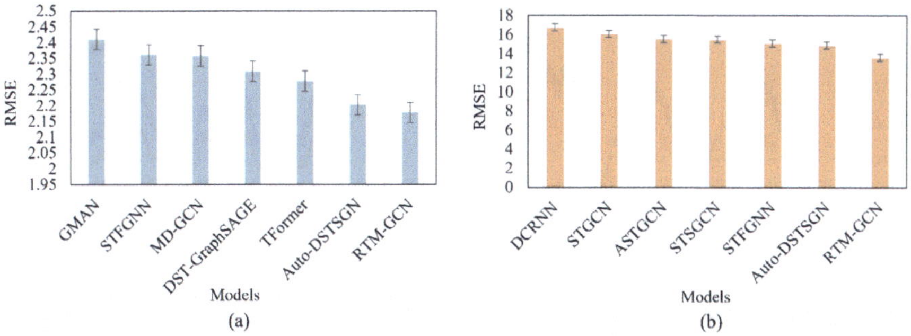

Fig. 2.44 Comparison of prediction performance under special events. (**a**) Comparison of prediction performance during peak hours. (**b**) Comparison of prediction performance during holiday periods

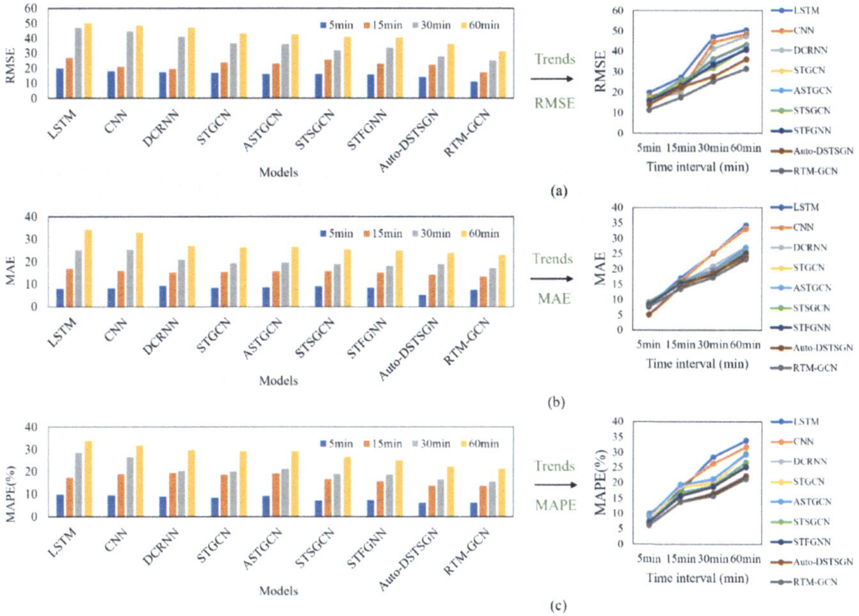

Fig. 2.45 Performance comparison of long-term time feature prediction. (**a**) RMSE. (**b**) MAE. (**c**) MAPE

increases, the predictive accuracy of all models decreases accordingly. However, RTM-GCN maintains the best predictive performance for most of the time and has excellent long-term feature extraction capabilities.

To verify the role of the multi-factor fusion matrix, the traffic flow data of sensors #317852 and #313487 in PeMS03 from November 17, 2018 to November 23, 2018 were selected for testing in this paper. Figure 2.46a shows that the traffic flow of the two sensors has the same overall trend although there are large differences

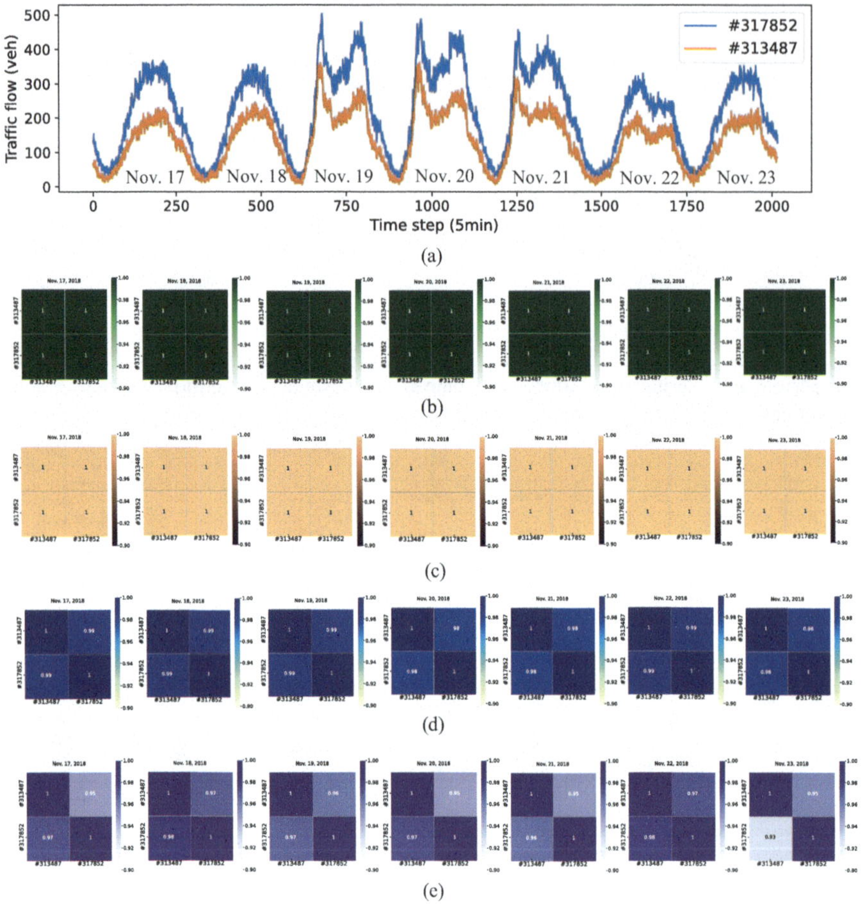

Fig. 2.46 Comparison of the results with the classical nodal relationship matrix. (**a**) Comparison of real traffic flow between sensors #317852 and #313487. (**b**) Adjacency matrix between sensors #317852 and #313487. (**c**) Distance matrix between sensors #317852 and #313487. (**d**) Data correlation matrix between sensors #317852 and #313487. (**e**) Multi-factor fusion feature between sensors #317852 and #313487

during the peak traffic period. This paper selects three types of relationship matrices to show the change of the relationship between these two nodes during the 7 days, as shown in Fig. 2.46b–d. It can be observed that the adjacency matrix and distance matrix do not change over time and cannot dynamically reflect traffic trends, while the data correlation matrix ignores the bidirectional traffic flow between nodes, although it may vary over time. The multi-factor fusion matrix changes with time as shown in Fig. 2.46e, which can better respond to the dynamic changes between nodes and also consider the bidirectional of traffic flow between nodes.

Ablation Experiments

To verify the importance of internal and external factor matrices for RTM-GCN, five variants were set up in this paper: RTM-GCN-NI (without internal factor matrix), RTM-GCN-NEW (without weather measurement matrix), RTM-GCN-NEN (without news measurement matrix), RTM-GCN-NE (without external factor matrix), and RTM-GCN-NIE (without both internal and external factor matrices, only adjacency matrix). To investigate whether the internal and external factor matrices have an impact on other models, they were introduced into STSGCN. The variant was named STSGCN-IE (considering both internal and external factor matrices). Table 2.26 shows the comparison results of several variants using PeMS07 as a test set. Compared with RTM-GCN, the spatial and temporal complexity of RTM-GCN-NIE decreased by 46.166% and 74.419%, respectively, while RMSE error accuracy increased by 12.139%. Compared with STSGCN-IE, STSGCN's spatial and temporal complexity decreased by 74.040% and 55.905%, respectively, while RMSE error accuracy increased by 6.143%. Therefore, it can be concluded that the proposed internal and external factor matrices have a significant impact on improving model accuracy.

To verify the scalability of the spatial enhancement module, this paper uses PeMS04 and PeMS08 as the test sets to demonstrate the MAE results of the seven models, as shown in Fig. 2.47. It can be found that by extending the spatial enhancement module, the prediction accuracies of the models in PeMS04 and PeMS08 are improved by 1.807% and 1.792%, respectively, which indicates the superior scalability of the spatial enhancement module.

The multi-factor graph convolution module in RTM-GCN extracts the local spatial features of the multi-factor matrix by setting multiple double-layer graph convolutions. To determine the effect of the number of double-layer graph convolutions on the prediction results, they are set to 1–5 in this paper, respectively. The test results of the four test sets are shown in Fig. 2.48. When the number of double-layer graph convolutions in PeMS (03–08) is set to 2, the average test results of the model are optimal.

Table 2.26 Impact of internal and external factor matrices

Model	Params(M)	FLOPs(G)	RMSE
STSGCN	**1.095**	32.582	38.894
STSGCN-IE	4.218	73.890	36.643
RTM-GCN-NIE	2.731	**10.682**	36.426
RTM-GCN-NE	3.512	21.079	34.606
RTM-GCN-NI	3.512	21.079	33. 643
RTM-GCN-NEW	3.573	31.318	33.367
RTM-GCN-NEN	4.293	31.317	32.914
RTM-GCN	5.073	41.757	**32.483**

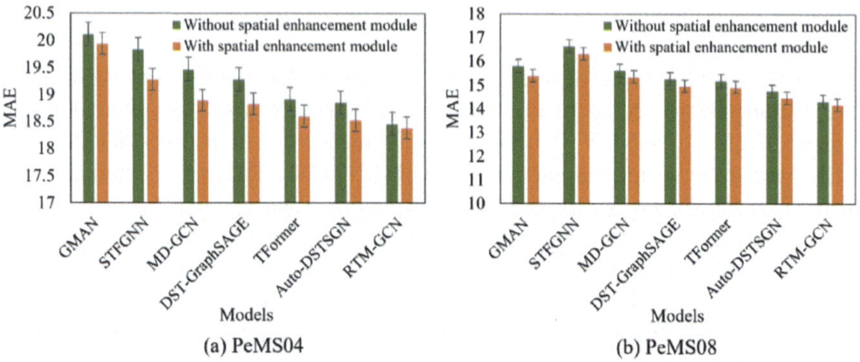

Fig. 2.47 Scalability analysis of spatial enhancement module. (**a**) PeMS04. (**b**) PeMS08

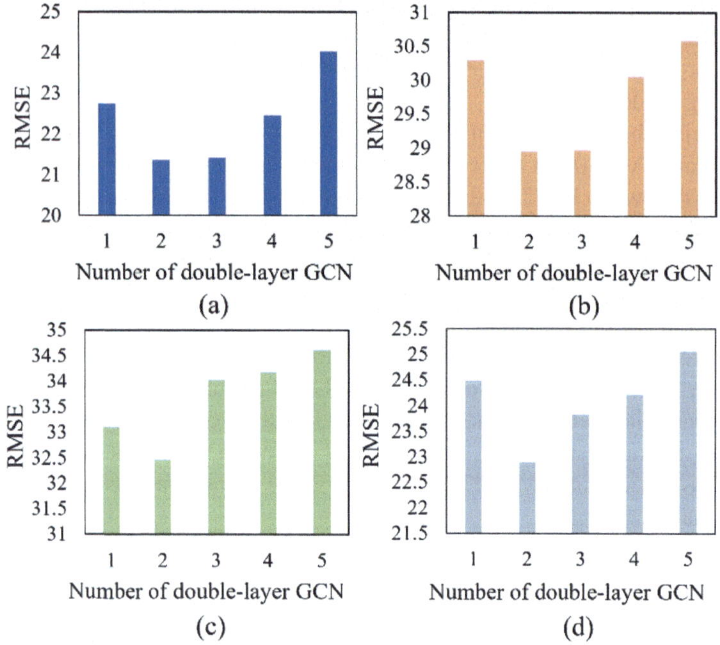

Fig. 2.48 Ablation experiment on the number of double-layer GCN. (**a**) PeMS03. (**b**) PeMS04. (**c**) PeMS07. (**d**) PeMS08

Due to the inevitable interference from noise received during the transmission of traffic flow, this paper uses perturbation analysis to test the noise immunity of the RTM-GCN to verify the robustness. In this paper, random noise with Gaussian distribution $N_{\text{Gau}} \in (0, \ \sigma^2)$ and gamma distribution $N_{\text{Gam}} \in (\varepsilon, \ 0.5)$ is added to the input data. Using PeMS08 as the test data, Fig. 2.49a, b show the effect of adding Gaussian noise and gamma noise to the dataset, respectively. From Fig. 2.49, it can be found that the prediction performance changes very little regardless of the noise,

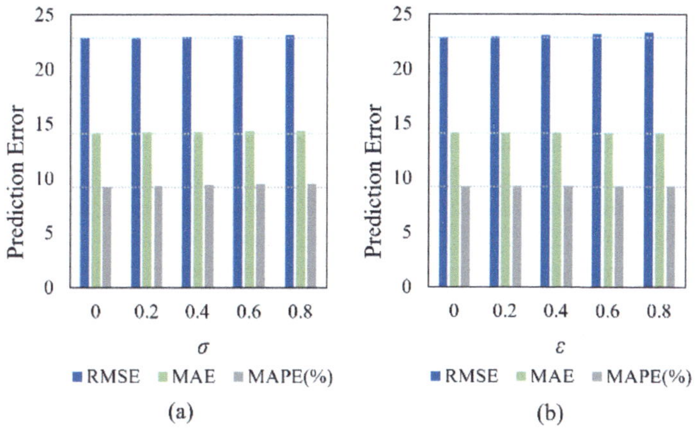

Fig. 2.49 Perturbation analysis. (**a**) Add Gaussian noise to input data. (**b**) Add gamma noise to input data

Fig. 2.50 The impact of simplified traffic raster features on model accuracy

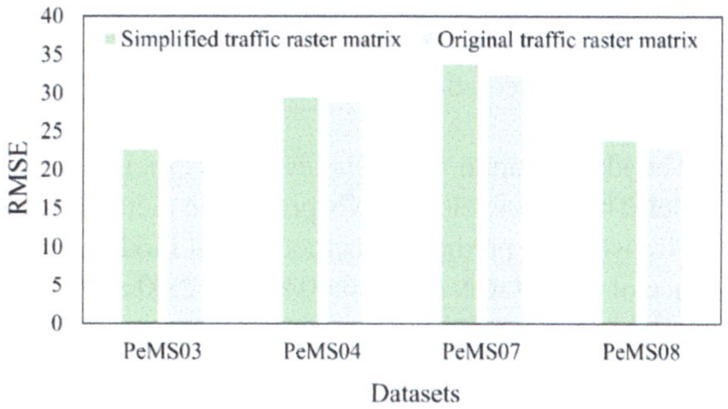

indicating that RTM-GCN has high robustness to achieve traffic flow prediction under high noise pollution.

For datasets without latitude and longitude information, this paper uses simplified traffic raster features to enhance global features. To verify the impact of simplified traffic raster features and original traffic raster features on model accuracy, the PeMS datasets are selected for testing. Using RMSE as the evaluation indicator, the specific results are shown in Fig. 2.50. Compared with the original traffic grid features, the simplified traffic grid features will lose some global features, leading to an inevitable decrease in model accuracy.

2.5.5.4 Discussion and Future Work

Accurately capturing traffic peaks and valleys is crucial for improving the accuracy of traffic flow prediction models. The PeMS (03–08) dataset exhibits obvious traffic peaks and valleys, with traffic volume gradually increasing from 4 am and reaching its first peak around 8:30 am, before declining again around 6 pm. Figure 2.41 shows that RTM-GCN has good fitting ability for traffic flows in different regions. Figure 2.42 demonstrates that RTM-GCN can achieve good convergence effects in multiple datasets, indicating its strong generalization capability. Special events have a significant impact on traffic flow, and this paper uses nodes in PeMS03 as an example to show how different holidays affect the periodic changes of traffic flow. Figure 2.43 verifies the sensitivity of RTM-GCN to special events, achieving accurate traffic flow predictions under different special event scenarios.

Table 2.25 demonstrates the results of RTM-GCN compared with three types of model-driven method, single feature extraction method, and GNN prediction method. Using RMSE as the evaluation metric, RTM-GCN improves the accuracy of the best-performing model in the baseline models by 6.362%. GNN prediction methods are the best performers among the three types in Table 2.25, among which Auto-DSTSGN is the best performer among the GNN prediction methods due to the fact that it not only synchronizes the extraction of spatial-temporal features, but also more flexibly inscribes spatial-temporal relationships at different time steps and even different scenarios through graph structure search. However, the RTM-GCN proposed in this paper not only improves the extraction of spatial-temporal features, but also introduces global feature enhancement at the same time, which improves the prediction performance of the model based on the fusion of multi-graph factors.

The identification of special events is crucial for the practical application of the model. This paper selects GNN prediction method, GraphSAGE prediction method, and transformer prediction method as test models to compare the prediction performance of RTM-GCN from 10:00 am to 12:00 am and special holidays. As shown in Fig. 2.44, RTM-GCN has superior performance for predicting special events with superior global and local feature extraction. To assess the long-term prediction capabilities of all models, this paper divides the prediction cycle into intervals of 5 min, 15 min, 30 min, and 60 min. As shown in Fig. 2.45, with an increase in the prediction cycle duration, the prediction accuracy of all models decreases. The RTM-GCN model proposed in this paper maintains superior predictive performance for most time periods. Figure 2.46 demonstrates that the multi-factor fusion feature introduced in this paper better captures the bidirectional variation features of traffic flow between nodes, in comparison to classical adjacency matrices, distance matrices, and data association matrices. Through ablation experiments, this paper not only verifies the importance and scalability of internal and external factor matrices to models but also confirms that spatial enhancement modules have a significant impact on improving model accuracy. Although both GraphSAGE and transformer prediction methods emphasize the enhancement of global or local features, RTM-GCN introduces multi-graph spatial feature extraction on top of keeping the global distribution unchanged in the spatial enhancement module, which effectively achieves improved prediction performance. The good robustness of RTM-GCN is proved by the perturbation analysis of Gaussian noise and gamma noise. Compared with the original traffic raster features, the simplified traffic raster features will lose some global features, leading to an inevitable decrease in model accuracy.

Although RTM-GCN shows good prediction performance on PeMS (03–08) and SZ-TAXI datasets, there are still some areas that need improvement: (1) the proposed RTM-GCN has a complex structure, which improves the prediction accuracy compared with baseline models such as STSGCN, but also leads to increased spatial and temporal complexity; (2) the proposed RTM-GCN in this paper can be used by attention mechanism to achieve the interpretability of local network connections, but it is difficult to meet the interpretability requirements of global network structure, and the whole neural network architecture is still a black box; (3) the convergence of the model depends more on the choice of optimizer and network structure, and it is difficult to verify whether the structure of the model is optimal through the derivation of mathematical theory. To address the above shortcomings, this paper will focus on improving the model structure, interpretability, and mathematical derivation in the future. (1) Optimization of the model structure through heuristic algorithms will be investigated to improve accuracy while ensuring that the spatial-temporal complexity of the model remains constant; (2) more attentional or diffusion models will be introduced into the overall network structure in order to visualize the outputs of each component and to improve the interpretation; and (3) the selection of optimizers will be investigated, and the optimal optimizers will be studied and selected by proving mathematical convergence.

2.6 Conclusion

In summary, this chapter highlights the significance of spatial-temporal graph convolutional networks in enhancing connected traffic prediction. By leveraging the synergy between static and dynamic spatial correlations, the discussed models provide robust frameworks for accurately capturing complex traffic patterns and their temporal dependencies. The exploration of various methodologies, including SDSCNN and residual attention-enhanced models, demonstrates their effectiveness in addressing challenges related to spatiotemporal heterogeneity. Overall, this chapter establishes a comprehensive foundation for understanding how STGCNs can significantly improve prediction accuracy and robustness in connected traffic systems.

References

1. T. Alghamdi, K. Elgazzar, M. Bayoumi, T. Sharaf, S. Shah, Forecasting traffic congestion using ARIMA modeling, in *2019 15th International Wireless Communications & Mobile Computing Conference (IWCMC), (Tangier, Morocco, 2019)*, pp. 1227–1232. https://doi.org/10.1109/IWCMC.2019.8766698
2. S.R. Chandra, H. Al-Deek, Predictions of freeway traffic speeds and volumes using vector autoregressive models. J. Intell. Transp. Syst. **13**(2), 53–72 (2009). https://doi.org/10.1080/15472450902858368

3. L. Zhao, Y. Song, C. Zhang, Y. Liu, P. Wang, T. Lin, M. Deng, H. Li, T-GCN: A temporal graph convolutional network for traffic prediction. IEEE Trans. Intell. Transp. Syst. **21**(9), 3848–3858 (2020). https://doi.org/10.1109/TITS.2019.2935152
4. Y. Li, R. Yu, C. Shahabi, Y. Liu, Diffusion convolutional recurrent neural network: Data-driven traffic forecasting. arXiv (2018) http://arxiv.org/abs/1707.01926. Accessed 19 Jul 2022
5. S. Guo, Y. Lin, N. Feng, C. Song, H. Wan, Attention based spatial-temporal graph convolutional networks for traffic flow forecasting. Proc. AAAI Conf. Artif. Intell. **33**(01) (2019). https://doi.org/10.1609/aaai.v33i01.3301922
6. Z. Wu, S. Pan, G. Long, J. Jiang, C. Zhang, Graph WaveNet for deep spatial-temporal graph modeling. arXiv (2019) http://arxiv.org/abs/1906.00121. Accessed 19 Jul 2022
7. Q. Zhang, J. Chang, G. Meng, S. Xiang, C. Pan, Spatio-temporal graph structure learning for traffic forecasting. Proc. AAAI Conf. Artif. Intell. **34**(01) Art. no. 01 (2020). https://doi.org/10.1609/aaai.v34i01.5470
8. Z. Zhang, Y. Li, H. Song, H. Dong, Multiple dynamic graph based traffic speed prediction method. Neurocomputing **461**, 109–117 (2021). https://doi.org/10.1016/j.neucom.2021.07.052
9. C. Song, Y. Lin, S. Guo, H. Wan, Spatial-temporal synchronous graph convolutional networks: a new framework for spatial-temporal network data forecasting, in: AAAI(2020), AAAI, Palo Alto, USA. pp. 914–921, (2020). https://doi.org/10.1609/aaai.v34i01.5438
10. P. Ramachandran, N. Parmar, A. Vaswani, I. Bello, A. Levskaya, J. Shlens, Stand-alone self-attention in vision models, in: NeurIPS 2019, Vancouver, Canada. pp. 1–13, (2019).

Chapter 3
Dynamic and Multi-Graph Approaches for Connected Traffic Flow Prediction

3.1 Complex Asymmetric Information in the Transportation Road Network

Traffic prediction is based on historical records of multiple nodes distributed in the road network to predict future traffic conditions. Nevertheless, as a result of the uneven configuration of the city's road network structure [1, 2], traffic flow data have complex temporal and spatial correlations [3, 4]. There are currently several issues in research in this area:

- Dynamic correlation between nodes. In previous studies, the correlation between nodes was often described using a static adjacency matrix, but in reality, the relationship among intersections within the traffic grid fluctuates over time. As shown in Fig. 3.1, residents departing from a residential area to work in an industrial area often choose the route marked with a blue arrow due to its shorter distance. However, if a traffic accident occurs on this section, rendering it impassable, residents would be forced to choose the route marked with a yellow arrow to reach the industrial area. During this process, the originally strong correlation between nodes 1 and 2 would weaken or even become irrelevant. Therefore, learning the dynamic spatiotemporal correlations between nodes is very necessary [5, 6].
- Oversmoothing problem of graphs. Although existing models have achieved relatively good results in traffic prediction using deep graph neural networks, deep graph neural networks gradually lose the graph structure and node feature information, leading to a decline in network performance. Therefore, it is necessary to forget irrelevant information layer by layer progressively.
- Long-term temporal feature extraction. Traffic data are not only influenced by short-term traffic conditions but also by long-term time dependence, such as people using a certain road to commute to work and returning home at the end of the day. On workdays, traffic congestion on the road significantly increases dur-

Q. Shi et al., *Connected Vehicles Traffic Prediction*, Wireless Networks,
https://doi.org/10.1007/978-3-031-84548-2_3

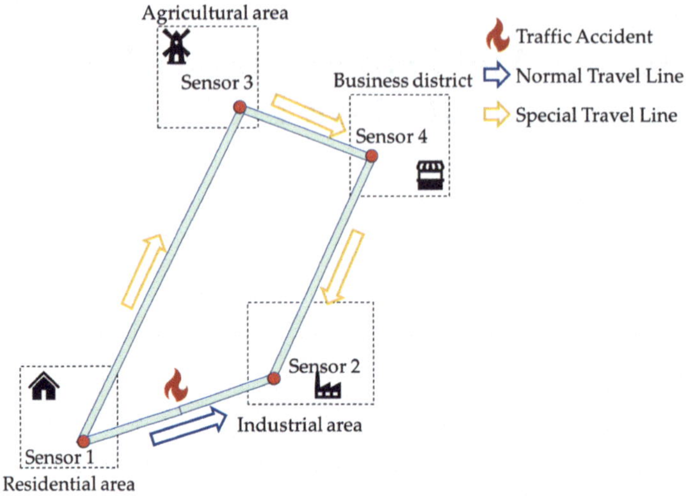

Fig. 3.1 Spatiotemporal correlation between road network nodes

ing morning and evening rush hours. However, on weekends or holidays, as most people do not go to work, the traffic flow on this road will be reduced and congestion will be significantly reduced. Therefore, observing the traffic data of this road over a long period will reveal clear cyclical changes in traffic flow and congestion.

3.2 Multi-Sequential Temporal Convolution Gated Graph Neural Network (MTCGGN)

In order to further explore the spatiotemporal characteristics of traffic flow length series, the MTCGGN model is proposed in this paper, and its overall architecture is shown in Fig. 3.2. The model consists of four main parts:

1. Short-term spatiotemporal feature extraction module is used for processing adjacent sequences. The module uses GCN to extract the spatial features of the road network and GRU to extract the temporal features of the adjacent sequences.
2. Multi-temporal spatiotemporal feature extraction module is used to process periodic sequences. The module uses MTCN to capture the period characteristics of long series historical data.
3. External feature extraction module is used to deal with external factors. The module uses unique thermal coding to characterize weather and holidays and sets two fully connected layers to extract external features.
4. Attention fusion output module is used to integrate spatiotemporal features and external factor features. The module first fuses the temporal and spatial features of adjacent sequences with the periodic features of long series historical data to

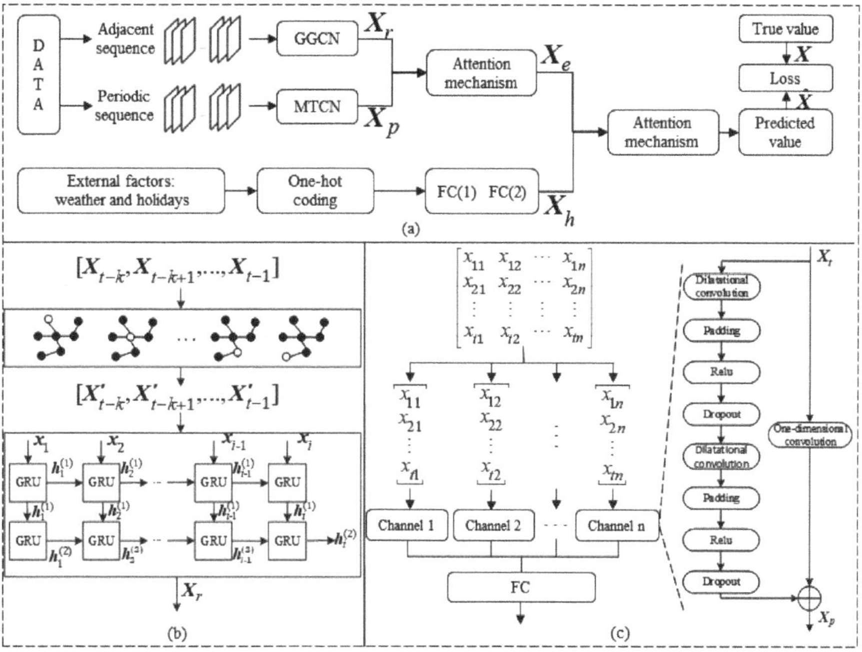

Fig. 3.2 Structure of MTCGGN. (**a**) MTCGGN network structure. (**b**) GGCN network structure. (**c**) MTCN network structure

obtain the temporal and spatial features of the road network, and then fuses external features to enhance the external environment perception ability of multi-temporal temporal features.

3.2.1 Sequence Partitioning

Assuming that the road network traffic flow data frame at time i is $R(i)$, the extraction rules of traffic flow dataset are as follows:

Adjacent sequences, $\text{Recent}(i) = \{R(i - T_h), \ldots, R(i - 3), R(i - 2), R(i - 1)\}$, T_h value is the length of the adjacent time series.

Period sequences, $\text{Period}(i) = \{R(i - T_d \times p), \ldots, R(i - 3 \times p), R(i - 2 \times p), R(i - 1 \times p)\}$, where p represents the span of a day and T_d represents the length of a sequence of historical periods. The specific meaning is to extract the data consistent with the moment where i moment is.

Extracting key frames instead of directly inputting all the data into the model can not only reduce the computational complexity of the model, but also highlight the time attribute of the data.

3.2.2 Short-Term Spatiotemporal Feature Extraction Module

As a neural network for processing topological data, graph convolutional network can directly map the original and real spatial attributes of the road network and can obtain the spatial features of the road network more accurately.

Thomas Kip [1] derived a new and most widely used form of GCN based on second-order Chebyshev polynomials.

$$\boldsymbol{H}^{(l+1)} = \sigma\left(\widetilde{\boldsymbol{D}}^{-\frac{1}{2}}\widetilde{\boldsymbol{A}}\widetilde{\boldsymbol{D}}^{-\frac{1}{2}}\boldsymbol{H}^{(l)}\boldsymbol{W}^{(l)}\right) \tag{3.1}$$

where $\boldsymbol{A} \in \mathrm{R}^{N\times N}$ denotes the adjacency matrix, $\widetilde{\boldsymbol{A}} = \boldsymbol{A} + \boldsymbol{I}$ denotes the adjacency matrix adding a self-connection, $\widetilde{\boldsymbol{D}}$ is the $\widetilde{\boldsymbol{A}}$ degree matrix; $\boldsymbol{H}^{(l)} \in \mathrm{R}^{N\times F}$ is the node eigenmatrix of layer l, $\boldsymbol{X} \in \mathrm{R}^{N\times F}$ representing the initial eigenmatrix of the node, N represents the number of nodes, F is the dimension of the eigenvector of each node, $\boldsymbol{H}^{(0)} = \boldsymbol{X}$; $\boldsymbol{W}^{l}$ represents the l layer weight and $\sigma(\cdot)$ is the Sigmoid activation function.

GRU is a common recurrent neural network variant, which absorbs the advantages of LSTM's long-term time feature extraction and improves computational efficiency. Let $\boldsymbol{x}_t$ represent the input at time t and $\boldsymbol{h}_t$ represent the output at time t, as defined in Eq. (3.2):

$$\boldsymbol{h}_t = (1 - z_t) \odot \boldsymbol{h}_{t-1} + z_t \odot \widetilde{\boldsymbol{h}}_t \tag{3.2}$$

Among them, operation symbol $\odot$ represents Hadamard product; z_t represents the updated gating signal, and its definition is shown in Eq. (3.3); $\boldsymbol{r}_t$ means reset gate signal, as defined in Eq. (3.4); $\widetilde{\boldsymbol{h}}_t$ represents the candidate hidden layer state, defined as shown in Eq. (3.5).

$$z_t = \sigma\left(\boldsymbol{W}_u \cdot [\boldsymbol{x}_t, \boldsymbol{h}_{t-1}] + \boldsymbol{b}_u\right) \tag{3.3}$$

$$\boldsymbol{r}_t = \sigma\left(\boldsymbol{W}_r \cdot [\boldsymbol{x}_t, \boldsymbol{h}_{t-1}] + \boldsymbol{b}_r\right) \tag{3.4}$$

$$\widetilde{\boldsymbol{h}}_t = \mathrm{Tanh}\left(\boldsymbol{W}_h \cdot [\boldsymbol{x}_t, \boldsymbol{r}_t \odot \boldsymbol{h}_{t-1}] + \boldsymbol{b}_h\right) \tag{3.5}$$

where $\boldsymbol{W}_u$, $\boldsymbol{W}_r$, $\boldsymbol{W}_h$ are the weight parameters of update gate, reset gate, and candidate hidden layer state, respectively; $\boldsymbol{b}_u$, $\boldsymbol{b}_r$, $\boldsymbol{b}_h$ are bias parameters of update gate, reset gate, and candidate hidden layer state, respectively; Tanh($\cdot$) is the Tanh activation function.

The Graph Convolutional Gated Unit Network (GGCN) structure is shown in Fig. 3.2b. $\boldsymbol{X} = [\boldsymbol{X}_{t-k}, \boldsymbol{X}_{t-k+1}, \ldots, \boldsymbol{X}_{t-1}]$ is the traffic flow data containing k continuous time segments, and the data is put into the convolutional network part of the Fig. 3.3 to obtain $[\boldsymbol{X}'_{t-k}, \boldsymbol{X}'_{t-k+1}, \ldots, \boldsymbol{X}'_{t-1}]$. Secondly, the time series data is put into the GRU for calculation, and the result $\boldsymbol{X}_r$ is obtained.

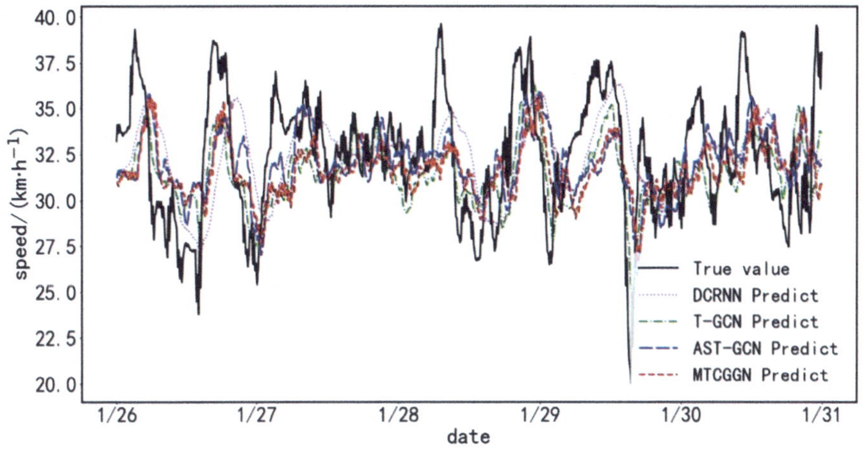

Fig. 3.3 Visualization of model predictions

3.2.3 Multi-Time Series Spatiotemporal Feature Extraction Module

The basic idea of MTCN is to stack multiple TCN in parallel, and each TCN processes the data of one input channel to learn the dependency relationship between multiple time series [2]. Compared with the traditional recurrent neural network model, TCN does not need to maintain long-term state information and can capture longer time dependence and avoid gradient explosion and gradient disappearance when processing long sequences. Its structure is shown in Fig. 3.2c. Taking time t as an example, the TCN calculation process for each channel is as follows:

Let the long historical sequence $\boldsymbol{X} = [\boldsymbol{X}_{t-k}, \boldsymbol{X}_{t-k+1},\ldots, \boldsymbol{X}_{t-1}]$, predict the future time $\boldsymbol{X}_t$, filter $\boldsymbol{\Phi} = [\varphi_1, \varphi_2,\ldots,\varphi_\tau]$, on the ι layer, the filter is defined as $\varphi_{\tau\iota}$, and the expansion rate at time $\boldsymbol{X}_t$ is d, then the convolution of expansion causality at time $\boldsymbol{X}_t$ is as follows:

$$\widehat{X}_t = \left(X *_d \varphi\right)_t = \sum_{\iota=1}^{I} \varphi_\iota X_{t-(I-\iota)d} \tag{3.6}$$

Thus, the size of the expansive convolution perception field is $(I-1)d + 1$, and the size of the perception field can be adjusted by increasing the size of I or d.

3.2.4 External Feature Extraction Module

One-hot encoding is used to convert non-numerical class data into numerical data [3]. For a set with Q states $\boldsymbol{Q} = \{Q_1, Q_2,\ldots, Q_q\}$, the encoded vector is 0 except for the index bit marked as 1.

Weather conditions are divided into five categories: sunny, cloudy, misty, light rain, and heavy rain, each of which corresponds to a unique thermal code. So the set of weather with 5 states is $\boldsymbol{Q} = \{Q_1, Q_2, \ldots, Q_5\}$, the encoding vector of the weather set $\boldsymbol{Q}$ is shown in Eq. (3.7).

$$\begin{cases} \text{sunny} = [1,0,0,0,0] \\ \text{cloudy} = [0,1,0,0,0] \\ \text{misty} = [0,0,1,0,0] \\ \text{light rain} = [0,0,0,1,0] \\ \text{heavy rain} = [0,0,0,0,1] \end{cases} \tag{3.7}$$

where $\widetilde{\boldsymbol{X}}$ is the result of the spatial correlation.

Holidays are divided into two categories: working days and holidays. Therefore, the encoding vector of holiday $\boldsymbol{Q}$ is shown in Eq. (3.8).

$$\begin{cases} \text{working days} = [1,0] \\ \text{holidays} = [0,1] \end{cases} \tag{3.8}$$

The input data of FC network includes weather and holiday information, and the external factors are characterized by one-hot coding, and the input data is normalized. Set up two fully connected layers, use the Sigmoid activation function, and extend the output value to the same size as $\boldsymbol{X}$.

3.2.5 *Attention Fusion Output Module*

The traffic flow of each detection point in the network is affected by the proximity and periodicity of the time attribute, but the degree of influence is different at different detection points. To this end, this paper uses the attention mechanism to integrate the short-term spatiotemporal feature extraction module and the multi-temporal spatiotemporal feature extraction module of the model. The equation is as follows:

$$\alpha_r = \frac{\exp(\boldsymbol{X}_r \boldsymbol{W}_r + b_r)}{\exp(\boldsymbol{X}_r \boldsymbol{W}_r + b_r) + \exp(\boldsymbol{X}_p \boldsymbol{W}_p + b_p)} \tag{3.9}$$

$$\alpha_p = \frac{\exp(\boldsymbol{X}_p \boldsymbol{W}_p + b_p)}{\exp(\boldsymbol{X}_r \boldsymbol{W}_r + b_r) + \exp(\boldsymbol{X}_p \boldsymbol{W}_p + b_p)} \tag{3.10}$$

$$\boldsymbol{X}_e = \alpha_r \cdot \boldsymbol{X}_r + \alpha_p \cdot \boldsymbol{X}_p \tag{3.11}$$

where α_r and α_p are the attention scores used by each detection point to characterize the degree to which proximity and periodicity affect it. $\boldsymbol{W}_r$, $\boldsymbol{W}_p$, b_r, and b_p are learnable weights and biases in linear changes.

In the same way, finally, the integrated spatiotemporal feature X_e and the output X_h of the external feature extraction module are fused, and the predicted value of road network traffic data is expressed as follows:

$$\widetilde{X} = \alpha_e \cdot X_e + \alpha_h \cdot X_h \quad (3.12)$$

where α_e and α_h are the attention scores used by each detection point to describe the degree of influence of temporal and spatial features and external features on it.

Log-Cosh loss function can better deal with outliers and noise in traffic flow data and improve the robustness and stability of the model. Therefore, this paper uses Log-Cosh loss function to measure the error between the real value X and the predicted value $\widehat{X}$ of the average speed of the road network detection points and uses Adam optimization algorithm to optimize the model parameters.

$$\log \ \text{Cosh} = \sum_{i=1}^{n} \log\left(\cosh\left(X - \widehat{X}\right)\right) \quad (3.13)$$

3.2.6 Experiments

3.2.6.1 Experiments Settings

Dataset

The road network data of Luohu District, Shenzhen City was used to conduct the experiment. The traffic dataset contained 156 road sections, and the average speed of each road was measured every 15 min, with a total of 2976 data for 31 days from January 1, 2015 to January 31, 2015.

Parameter Setting

The experimental program was built based on the TensorFlow deep learning framework, and the dataset was processed and input into the model for training. Using K-fold cross-validation as dataset partitioning method, the hyperparameter cross-validation after grid search optimization is repeated 10 times. Parameter Settings: Batch size 64, training rounds 3000, learning rate 0.001, number of GCN layers 3, number of GRU cells 64, MTCN expansion rates 1, 2, 4, and 8.

Evaluation Indicators

The three indicators such as root mean square error (RMSE), coefficient of determination (R^2), and the mean absolute error (MAE) are all used to evaluate and quantify the prediction accuracy of the model, and the error between the predicted value and the true value is calculated to reflect the predictive performance of the model, which is defined as follows:

Mean Absolute Error (MAE)

$$\mathrm{MAE} = \frac{1}{n}\sum_{i=1}^{n}\left|y_i - \hat{y}_i\right| \tag{3.14}$$

Root Mean Square Error (RMSE)

$$\mathrm{RMSE} = \sqrt{\frac{1}{n}\sum_{i=1}^{n}\left(y_i - \hat{y}_i\right)^2} \tag{3.15}$$

Coefficient of Determination (R^2)

$$\mathrm{R}^2 = 1 - \frac{\sum_{i=1}^{n}\left(y_i - \hat{y}_i\right)^2}{\sum_{i=1}^{n}\left(y_i - \overline{y}_i\right)^2} \tag{3.16}$$

Among them, y_i represents the true value and $\hat{y}_i$ represents the predicted value.

3.2.6.2 Comparative Experimental Analysis

Single-Step Prediction

The prediction results of the MTCGGN model on the dataset were compared with six benchmark models, HA, SVR, GRU, DCRNN, T-GCN, and AST-GCN. Table 3.1 shows the performance of all models on the test set (single-step prediction for the next 15 min).

As shown in Table 3.1, the MTCGGN model proposed in this paper has better prediction performance than the other six models, and the three evaluation indicators all show that the MTCGGN model has the most accurate prediction results. Taking RMSE as the evaluation index, MTCGGN model is 2.485%,

Table 3.1 Model prediction results

Model	MAE	RMSE	R^2
HA	7.947	10.872	0.817
SVR	4.815	7.459	0.831
GRU	3.027	5.481	0.832
DCRNN	3.380	4.711	0.846
T-GCN	2.928	4.417	0.838
AST-GCN	2.933	4.305	0.850
MTCGGN	**2.814**	**4.198**	**0.859**

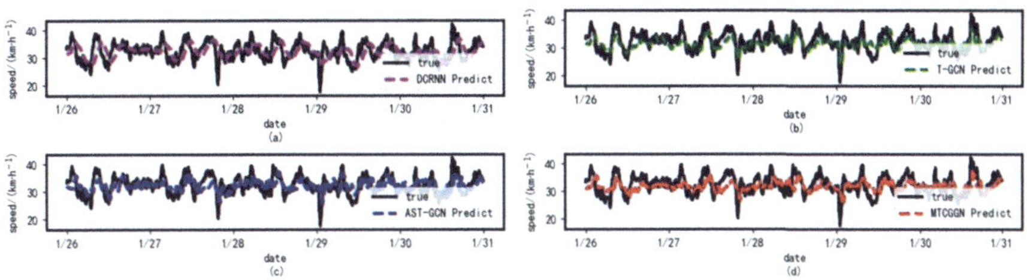

Fig. 3.4 A graph comparing the predicted value of a single model to the true value. (**a**) DCRNN model. (**b**) T-GCN model. (**c**) AST-GCN model. (**d**) MTCGGN model

4.958%, and 10.889% lower than DCRNN, T-GCN, and AST-GCN models, respectively.

The experimental results also show that due to the limited modeling ability of traditional models for complex traffic conditions in a large range of urban road networks, deep learning models have stronger expression ability than traditional models such as HA and SVR. The prediction accuracy of temporal models SVR and GRU is affected because they ignore the spatial characteristics of traffic flow. Although DCRNN and T-GCN models comprehensively consider the temporal and spatial characteristics of the road network, they do not consider external factors such as weather, resulting in low prediction accuracy. The AST-GCN model integrates weather and other external factors to improve the accuracy of traffic prediction, but it cannot fully tap the temporal and spatial periodicity of long series data. In contrast, MTCGGN model deals with adjacent sequence and periodic sequence, respectively, which can dig the historical cycle time characteristics hidden in traffic data more deeply and has good prediction accuracy.

In order to more intuitively compare the prediction performance of the model based on spatiotemporal feature mining, DCRNN, T-GCN, AST-GCN, and the MTCGGN model proposed in this paper were used to predict the road section No. 90217 in the road network. The visualization results of the model prediction from January 26 to January 31, 2015 are shown in Fig. 3.3.

It can be intuitively seen from the visualization results that MTCGGN has the best data fitting effect among the four models, and the predicted results of the model are most similar to the change trend of the real speed.

DCRNN, T-GCN, AST-GCN, and MTCGGN models were used to forecast the road section No. 90217 in the road network, respectively. The comparison between the predicted value of the model and the real value from January 26 to January 31, 2015 is shown in Fig. 3.4.

As seen in Fig. 3.4, the MTCGGN model is able to detect the beginning and end of the rush hour and has a similar trend to the actual traffic speed prediction results. However, for capturing the turning point of speed change, the MTCGGN model has a large deviation at the high and low peaks, indicating that the sudden change of traffic state is not only affected by the weather and holiday factors used in this paper, but also by other factors such as traffic accidents.

Multi-Step Prediction

In order to further test the prediction effect of the model, multi-step prediction is carried out in this paper, that is, the prediction results of the previous step are used to predict the next step, and the performance of the four models is compared with the increase of the prediction step length. Taking 15 min as a single time step, the predicted step length increased from 15 to 60 min, and the experimental results were shown in Fig. 3.5.

Figure 3.5 shows that with the gradual increase of the forecasting time, the forecasting difficulty becomes more and more serious, and the overall model error shows an upward trend. Compared with other models, MTCGGN in this paper has achieved better prediction results in the medium and long term. With the increase of prediction time, the prediction error of MTCGGN model grows more slowly than that of AST-GCN model. This is because MTCGGN model integrates the proximity characteristics and historical cycle characteristics of traffic flow dataset and mining the time dimension information of data more fully, thus showing greater advantages in medium- and long-term prediction.

Ablation Experiment Analysis

To further analyze the performance of MTCGGN model, different module combinations are used as follows: (1) MTCGGN_ggcn contains only periodic feature extraction module and external factor module; (2) MTCGGN_mtcn contains only spatiotemporal feature extraction module and external factor module; (3) MTCGGN_fc contains only spatiotemporal feature extraction module and periodic feature extraction module. These three variant models were compared with MTCGGN for ablation experiments.

As shown in Fig. 3.6, MAE and RMSE values of MTCGGN are smaller than those of MTCGGN_ggcn, MTCGGN_mtcn and MTCGGN_fc, which indicates that the spatiotemporal extraction module, cycle extraction module, and external factor extraction module can better extract the features of traffic flow and external factors.

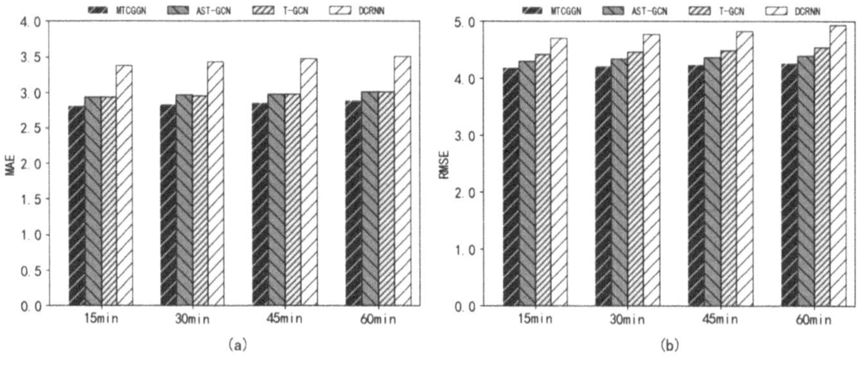

Fig. 3.5 Multi-step prediction MAE and RMSE bar charts. (**a**) Multi-step prediction MAE. (**b**) Multi-step prediction RMSE

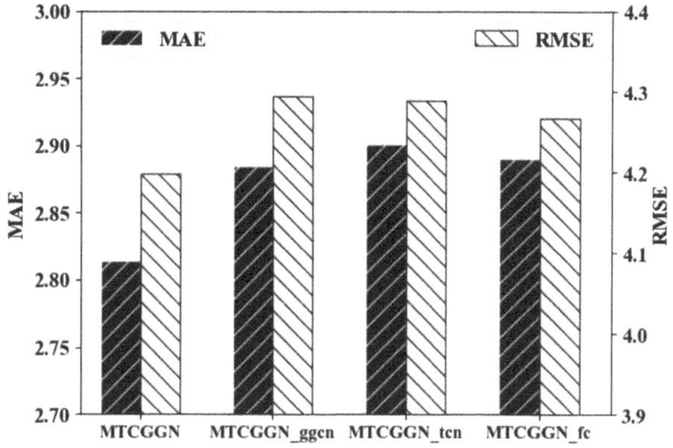

Fig. 3.6 Histogram of ablation experiment

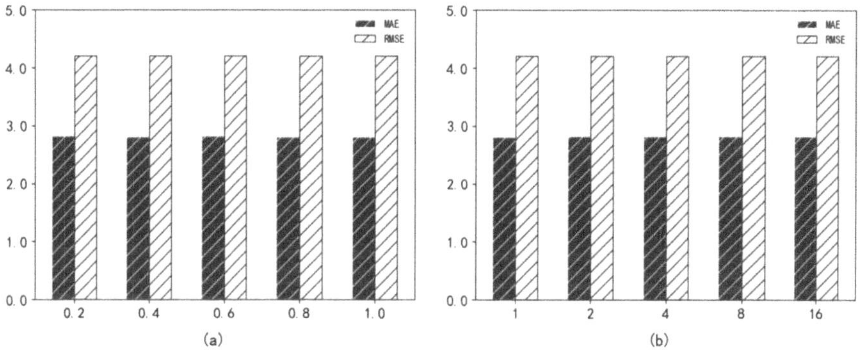

Fig. 3.7 Perturbation experiment results diagram. (**a**) Gaussian noise. (**b**) Poisson noise

Analysis of Disturbance Experiment

To verify the robustness of the model, Gaussian noise subject to Gaussian distribution $N \in (0, s^2)$, $s \in (0.2, 0.4, 0.6, 0.8, 1)$ and Poisson noise subject to Poisson distribution $P(\lambda)$, $\lambda \in (1, 2, 4, 8, 16)$ are injected into the data, respectively. Figure 3.7 shows the disturbance experiment results. Both MAE and RMSE of the MTCGGN model remain relatively stable under different noise settings, which verifies the robustness of the MTCGGN model.

3.3 Dynamic Spatiotemporal Correlation Graph Convolutional Network (MSGSGCN)

3.3.1 Overview of MSGSGCN

The model architecture proposed in our study is illustrated in Fig. 3.8, which includes an NCE module designed to evaluate the connectivity between nodes and multiple spatiotemporal modules aimed at capturing the spatiotemporal features of

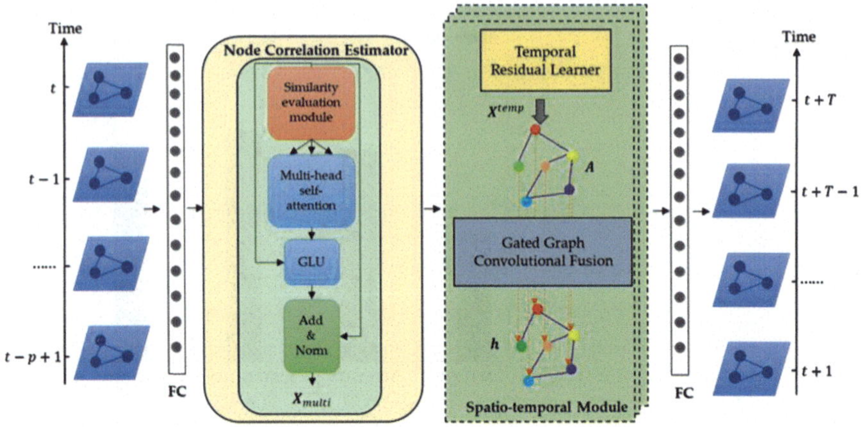

Fig. 3.8 Overall architecture

traffic data. Each of these modules is equipped with a TRL module for identifying long-range temporal relationships, and a GGCF module for the learning of spatial information and the integration of spatiotemporal data.

3.3.2 Node Correlation Estimator

There is a spatiotemporal correlation among different nodes in a road network. It is crucial to learn the correlation between each node and explain its causal relationship for accurate traffic prediction. For instance, Dynamic Time Warping (DTW) measures the similarity between two time series by calculating the minimum distance between them, but it has low computational efficiency and suboptimal performance.

While the standard self-attention module can also calculate the spatiotemporal correlation between nodes, it faces the issue of sparse attention distribution, which often leads to the model overlooking some important information. Therefore, in the NCE model, by integrating a similarity evaluation module that is primarily composed of dot product operations, the ability of the model to learn key information can be enhanced, avoiding an overly concentrated attention distribution on certain nodes.

The model first processes the input signal $\boldsymbol{X} \in \mathrm{R}^{N \times P \times F}$ through a linear layer and smooths the two linear processing results with BatchNorm and ELU activation functions. Then, it passes them through a separate linear layer, respectively, and multiplies the resulting matrices in parallel, activating the result with LeakyRelu. The multiplied result is then input into the multi-head self-attention module for further processing. In the multi-head self-attention module, three subspaces can be generated, including Query vector $\boldsymbol{Q}$, Key vector $\boldsymbol{K}$, and Value vector $\boldsymbol{V}$. The process of learning peripheral node features can be expressed as follows:

$$Q = XW^Q, \quad K = XW^K, \quad V = XW^V \tag{3.17}$$

where the relevance is calculated by computing Q and K for the dot product, and softmax is used to calculate the attention weight:

$$\text{Attention}(Q, K, V) = \text{softmax}\left(\frac{QK^T}{\sqrt{d_k}}\right)V \tag{3.18}$$

where d_k represents the dimension size of K. This division is called scaling, and it helps to stabilize the gradients during model training.

To enhance the fitting performance, this paper further employs multi-head self-attention, concatenating individual attention results to obtain the following:

$$X_{\text{multi}} = \text{Concat}(\text{head}_1, \ldots, \text{head}_h)W_a \tag{3.19}$$

where $\text{head}_i = \text{Attention}(Q_i, K_i, V_i)$ resents the attention weight calculation results of different heads. When using more heads to jointly focus on inputs from different representation subspaces, the model can learn more diverse latent information.

After using multi-head self-attention, the results are normalized by a layer norm, and the learning results of node correlation are multiplied with the original input Hadamard through the linear layer, which is equivalent to adding the embedding of node correlation to the original input and constructing the residual structure with the original input. Figure 3.9 shows the structure of the node correlation estimator.

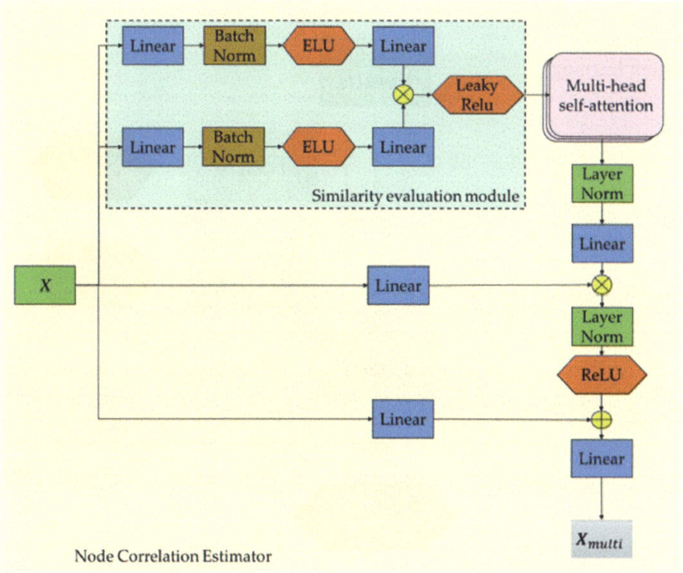

Fig. 3.9 Extraction of dynamic spatiotemporal correlations in road networks

3.3.3 Temporal Residual Learner

One-dimensional temporal convolution makes it easy to forget important information when dealing with long-term time dependencies. RNN can capture long-term dependency but has large computational consumption and is difficult to parallelize. Compared with the traditional model, the convolution method used by TCN is dilated causal convolution. The improved model has the advantages of parallel computing, long-term dependence modeling, stable gradient propagation, and flexibility in processing time series data, which can flexibly and effectively learn long-term time series features.

Although the TCN has advantages in extracting long-term temporal features, a single TCN may find it difficult to capture subtle distinctions in long-term temporal dependencies. However, combining multiple TCNs can significantly enhance the model's learning ability, thereby capturing features and patterns at multiple levels of the data. The overall framework is shown in Fig. 3.10.

The expression for the temporal residual learner can be expressed as follows:

$$\boldsymbol{X}^{\text{filter}} = \text{Tconv}\left(\boldsymbol{X}_{\text{multi}}\right) \tag{3.20}$$

$$\boldsymbol{X}^{\text{gate}} = \text{Tconv}\left(\boldsymbol{X}_{\text{multi}}\right) \tag{3.21}$$

$$\boldsymbol{X}^{\text{res}} = \text{Tconv}\left(\boldsymbol{X}_{\text{multi}}\right) \tag{3.22}$$

$$\boldsymbol{X}^{\text{temp}} = \text{LeakyReLU}\left(\boldsymbol{X}^{\text{res}} + \text{ELU}\left(\boldsymbol{X}^{\text{filter}}\right) \otimes \text{ELU}\left(\boldsymbol{X}^{\text{gate}}\right)\right) \tag{3.23}$$

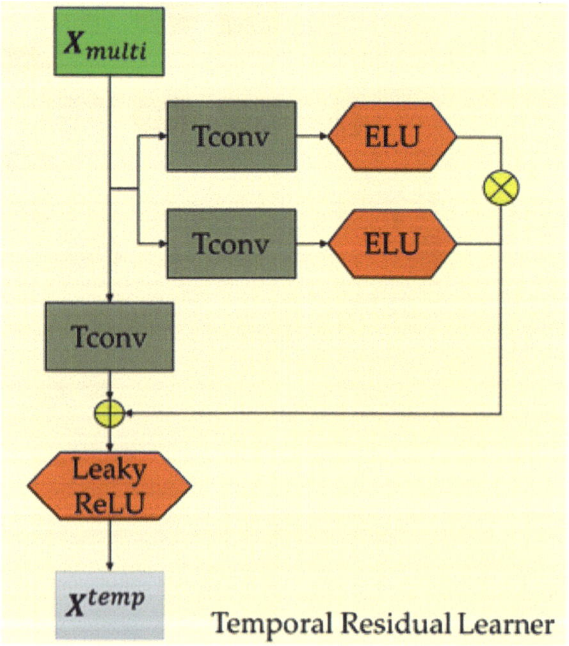

Fig. 3.10 Long-term temporal pattern learning

where Tconv denotes the dilated causal convolution operation, and ⊗ and denotes the Hadamard product. Compared to a single TCN module, the TRL module can more effectively learn time features of different layers and models through residual connections of multiple TCNs and reduce gradient attenuation to enhance the model's ability to learn long-term time features.

3.3.4 Gated Graph Convolutional Fusion

Although deep GCN networks have a stronger capability in extracting spatial features and can integrate more information between distant nodes, capturing long-distance dependencies between nodes, they suffer from the oversmoothing problem, where with increasing layers, different nodes tend to exhibit similar features, losing distinctiveness. Therefore, this paper employs a recurrent gating mechanism to control the propagation of information in deep GCNs by discarding irrelevant information layer by layer to enhance the model's ability to mine deep spatial features. This approach effectively avoids excessive dependencies between layers compared to ordinary residual methods.

As shown in Fig. 3.11, this paper uses a gating structure similar to GRU in the process of information transmission. By gating the control input and the processing results of the adaptive graph diffusion network, GCN can reduce the transmission of useless information and improve the calculation accuracy in the process of information transmission.

The expression of gated graph convolutional fusion can be expressed as follows:

$$z = \sigma\left(W_{zx} X^{\text{temp}} + W_{zg}\left[\text{GCN}\left(X^{\text{temp}}, A\right)\right]\right) \tag{3.24}$$

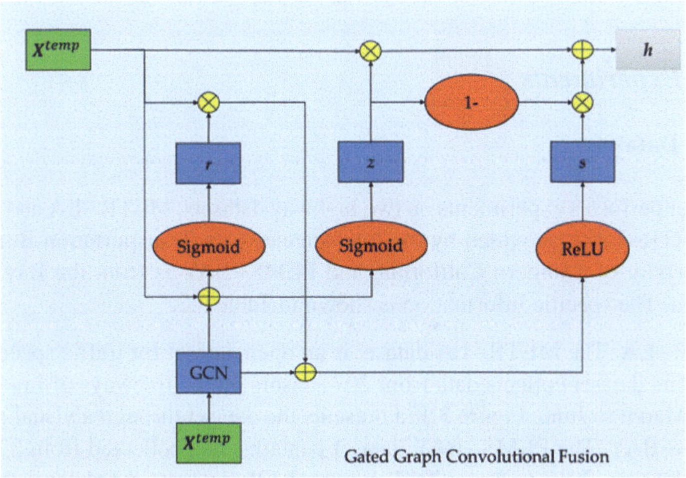

Fig. 3.11 Spatiotemporal feature fusion

$$\boldsymbol{r} = \sigma\left(\boldsymbol{W}_{rx}\boldsymbol{X}^{\text{temp}} + \boldsymbol{W}_{rg}\left[\text{GCN}\left(\boldsymbol{X}^{\text{temp}}, \boldsymbol{A}\right)\right]\right) \tag{3.25}$$

$$\boldsymbol{s} = \text{ReLU}\left(\boldsymbol{W}_{sx}\left[\text{GCN}\left(\boldsymbol{X}^{\text{temp}}, \boldsymbol{A}\right)\right] + \boldsymbol{W}_{sg}\left(\boldsymbol{r}^{*}\boldsymbol{X}^{\text{temp}}\right)\right) \tag{3.26}$$

$$\boldsymbol{h} = \boldsymbol{z}^{*}\boldsymbol{X}^{\text{temp}} + (1 - \boldsymbol{z}) * \boldsymbol{s} \tag{3.27}$$

where $\boldsymbol{X}^{\text{temp}}$ denotes the processing result of the above TRL module, and $[\text{GCN}(\boldsymbol{X}^{\text{temp}}, \boldsymbol{A})]$ denotes the adaptive diffusion graph convolution operation. σ is the activation function Sigmoid.

Two gating units, $\boldsymbol{z}$, and $\boldsymbol{r}$, activated with Sigmoid, are used to control how the previous information is updated at the current time and to control whether the previous information should be ignored at the current time, respectively. Controlling the retention and update of information can help GCNs process spatial information faster and more efficiently, avoiding the over-smoothing issue caused by multiple graph convolutions and enabling the model to simulate the spatial structure of real road networks.

3.3.5 *Loss Function*

Use Mean Absolute Error (MAE) as the loss function, where m represents the number of samples, y_i and $\hat{y}_i$ represent the actual value and the model predicted value, respectively.

$$\text{Loss} = \frac{1}{m}\left(\sum_{i=1}^{m}\left|y_i - \hat{y}_i\right|\right) \tag{3.28}$$

3.3.6 *Experiments*

3.3.6.1 Datasets

This paper performs experiments on two highway datasets, METR–LA and PEMS–BAY. METR–LA is provided by the Los Angeles City Transportation Bureau and the University of Southern California, and PEMS–BAY is from the Bay Area of California. The specific information is shown in Table 3.2.

- METR–LA. The METR–LA dataset is an open dataset for traffic speed prediction. The dataset collects data from 207 sensors on the freeways of Los Angeles from March to June. Figure 3.12a presents the dataset through a visual graph.
- PEMS–BAY. The PEMS–BAY dataset contains data collected from 325 nodes from January 2017 to June 2017. Figure 3.12b presents the dataset through a visual graph.

Table 3.2 Parameter settings

Dataset	METR–LA	PEMS–BAY
Area	Los Angeles	Bay Area of California
Nodes	207	325
Time interval	5 min	5 min
Target	Speed	Speed
Start time	1 March 2012	1 January 2017
End time	27 June 2012	30 June 2017

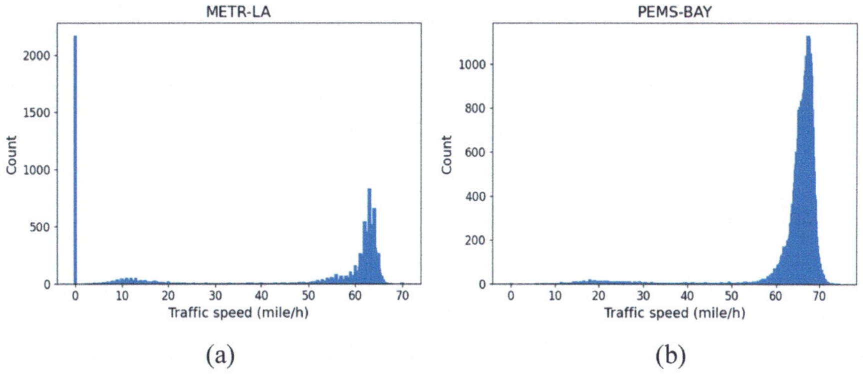

Fig. 3.12 Traffic speed distribution in the dataset. (**a**) METR-LA. (**b**) PEMS-BAY

3.3.6.2 Experimental Settings

Hyperparameter

In the experiments, Pytorch 1.6.1 was used to implement the experiments on a graphics card with 8 GB memory; the graphics card used NVIDIA GeForce RTX 3070, the batch size was set to 64, Adam was used to optimize, and the learning rate was 0.001. The dropout rate was set to 0.3 to prevent overfitting. Note that the number of heads in the METR–LA dataset and PEMS–BAY dataset was set to 4.

Baseline Models and State-of-the-Art Models

Eight traffic-prediction models are selected as baseline models to compare with the proposed model. In the experiments comparing with these baseline models, this paper references the parameter settings described in the baseline papers and relevant literature.

- ARIMA. Integrating moving-average autoregressive models and using the difference to deal with time series problems.
- SVR. Support vector regression, a commonly used time series analysis model.

- DCRNN. Diffusion convolutional recurrent neural networks that learn spatio-temporal features using diffusion convolutions.
- STGCN. Spatiotemporal convolutional models, combining graph convolutional layers and convolutional sequences to learn spatiotemporal features.
- ASTGCN. Time is divided into three parts: adjacent, daily, and weekly.
- STSGCN. Three consecutive adjacent time slices are constructed into a local spatial graph.
- CCRNN. A hierarchical coupling mechanism is proposed to fuse the adjacency matrices of different layers.
- ADN-FA. The model is based on an Attention Diffusion Network (ADN), which is a decoder–encoder architecture and does not require any prior knowledge. ADN-FA improves upon the original dot-product attention (DA) in the ADN model by replacing it with Fast Linear Attention (FA).

Evaluating Indicator

This study used four error metrics to evaluate the prediction effect of MSGSGCN. These are MAE, Mean Absolute Percentage Error (MAPE), Root Mean Square Error (RMSE), and Mean Hassanat Distance (MHD). Additionally, this study introduces the MHD as an evaluation metric in the ablation experiments, as it is less sensitive to outliers and better able to accurately assess model performance, where an MHD result closer to 0 indicates a more ideal outcome. It should be noted that the dataset used in this paper consists of vehicle speeds, which are all non-negative values. Therefore, a portion of the MHD formula used in the paper is specifically designed for non-negative values. The equations for MAPE, RMSE, and MHD are as follows:

$$\text{MAPE} = \frac{1}{m}\left(\sum_{i=1}^{m}\left|\frac{y_i - \hat{y}_i}{y_i}\right|\right) \tag{3.29}$$

$$\text{RMSE} = \sqrt[2]{\frac{1}{m}\left(\sum_{i=1}^{m}\left(y_i - \hat{y}_i\right)^2\right)} \tag{3.30}$$

$$\text{MHD} = \begin{cases} \dfrac{\sum\left(1 - \dfrac{1+\min\left(y_i,\hat{y}_i\right)}{1+\max\left(y_i,\hat{y}_i\right)}\right)}{m}, & \min\left(y_i,\hat{y}_i\right) \geq 0 \\ \dfrac{\sum\left(1 - \dfrac{1+\min\left(y_i,\hat{y}_i\right)+\left|\min\left(y_i,\hat{y}_i\right)\right|}{1+\max\left(y_i,\hat{y}_i\right)+\left|\min\left(y_i,\hat{y}_i\right)\right|}\right)}{m}, & \min\left(y_i,\hat{y}_i\right) < 0 \end{cases} \tag{3.31}$$

3.3.6.3 Experimental Results

Prediction Effect of MSGSGCN

Tables 3.3 and 3.4 display the comparative results of the model on two datasets.

To better evaluate the performance of the model, this paper includes a range of baseline models: a statistical model ARIMA; a machine model SVR; classic spatiotemporal models like DCRNN, STGCN, ASTGCN, STSGCN; and recently popular encoder–decoder architecture models CCRNN and ADN-FA. The variety of baseline models helps us better assess our model's improvements.

Using RMSE as the evaluation metric, on the METR–LA and PEMS–BAY datasets, MSGSGCN reduces the error by 45.2% and 30.3%, respectively, compared to the statistical model ARIMA, and by 47.3% and 36.0%, respectively, compared to the machine model SVR. Compared with classic spatiotemporal graph neural networks such as DCRNN, STGCN, and STSGCN, the reduction is 4.6%, 22.9%, and 37.8%, respectively, while on the PEMS–BAY dataset, the reduction is

Tables 3.3 Comparison experiments on the METR–LA dataset

Data	Models	15 min			30 min			60 min		
		MAE	RMSE	MAPE (%)	MAE	RMSE	MAPE (%)	MAE	RMSE	MAPE (%)
METR–LA	ARIMA	3.99	8.21	9.60	5.15	10.45	12.70	6.90	13.23	17.40
	SVR	3.39	8.45	9.30	5.05	10.87	12.10	6.72	13.76	16.70
	DCRNN	2.77	5.38	7.30	3.15	6.45	8.80	3.60	7.60	10.50
	STGCN	2.88	5.74	7.62	3.47	7.24	9.57	4.59	9.40	12.70
	ASTGCN	4.86	9.27	9.21	5.43	10.61	10.13	6.51	12.52	11.64
	STSGCN	3.31	7.62	8.06	4.13	9.77	10.29	5.06	11.66	12.91
	CCRNN	2.85	5.54	7.50	3.24	6.54	8.90	3.73	7.65	10.59
	ADN-FA	3.02	6.01	8.20	3.56	7.30	10.22	4.31	8.70	12.61
	MSGSGCN	2.78	5.36	7.26	3.14	6.32	8.67	3.54	7.25	10.16

Tables 3.4 Comparison experiments on the PEMS-BAY dataset

Data	Models	15 min			30 min			60 min		
		MAE	RMSE	MAPE (%)	MAE	RMSE	MAPE (%)	MAE	RMSE	MAPE (%)
PEMS-BAY	ARIMA	1.62	3.30	3.50	2.33	4.76	5.40	3.38	6.50	8.30
	SVR	1.85	3.59	3.80	2.48	5.18	5.50	3.28	7.08	8.00
	DCRNN	1.38	2.95	2.90	1.74	3.97	3.90	2.07	4.74	4.90
	STGCN	1.36	2.96	2.90	1.81	4.27	4.17	2.49	5.69	5.79
	ASTGCN	1.52	3.13	3.22	2.01	4.27	4.48	2.61	5.42	6.00
	STSGCN	1.44	3.01	3.04	1.83	4.18	4.17	2.26	5.21	5.40
	CCRNN	1.38	2.90	2.90	1.74	3.87	3.90	2.07	4.65	4.87
	ADN-FA	1.48	3.04	3.05	1.87	4.12	4.16	2.34	5.22	5.72
	MSGSGCN	1.33	2.84	2.80	1.67	3.81	3.77	1.97	4.53	4.61

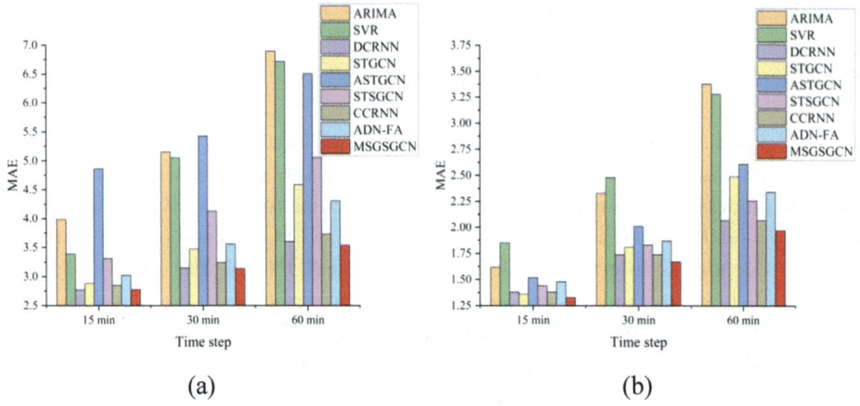

Fig. 3.13 Comparison with baseline models in terms of MAE. (**a**) METR-LA. (**b**) PEMS-BAY

4.4%, 20.4%, and 13.1%, respectively. Additionally, compared to the encoder–decoder structure-based ADN-FA model, there is also a reduction in error of 16.7% and 13.2%. Furthermore, Fig. 3.13 presents the visualization comparison of model performance for prediction step lengths of 15, 30, and 60 min.

In Fig. 3.13, it can be seen that as the prediction time steps increase, the prediction difficulty also increases, and the prediction errors of all models show an upward trend. However, the MSGSGCN still performs well in medium- and long-term prediction performance. Additionally, the experimental results also show that due to the limited processing ability of traditional machine models in handling realistic and complex traffic data, deep learning models generally outperform machine learning models, indicating that deep learning has stronger modeling ability in traffic speed prediction tasks. In addition, decoder–encoder-based models perform well in long-range prediction tasks, such as predicting for a 60 min time horizon. This is because this structure exhibits good parallelism, which provides significant advantages in handling longer time series data.

Visualization

To provide a more intuitive understanding and evaluation of the model, Fig. 3.14 features a well-crafted chart that visualizes the correlation between the model's predicted results and the actual data points. To demonstrate the model's generalization capability, we randomly selected two nodes from each dataset (nodes 5 and 16 from the METR–LA dataset and nodes 8 and 66 from the PEMS–BAY dataset) and depicted their comparative data across different time steps, showcasing the precision of the model's predictions. The model fits the real data well and has a robust predictive ability for peaks and troughs in traffic, indicating that the model can effectively forecast variations in traffic data. Notably, the model does not attempt to overfit some noise in the data. For instance, in the METR–LA dataset, there are some outliers present in the original data; hence, outlier values were replaced with

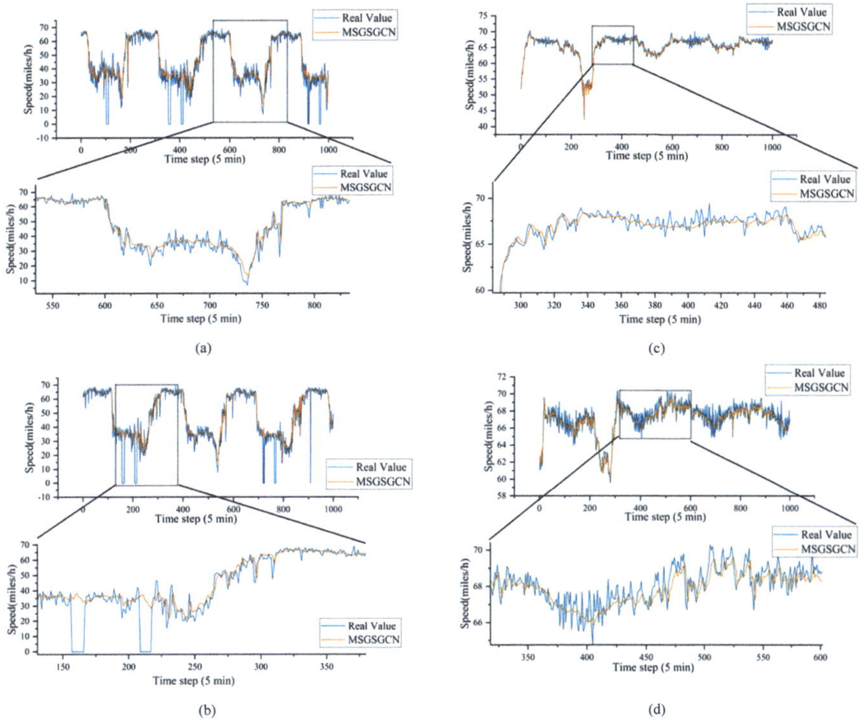

Fig. 3.14 Visualization of the METR–LA and PEMS–BAY datasets. (**a**) True vs. predicted values for node 5 in METR-LA dataset. (**b**) True vs. predicted values for node 16 in METR-LA dataset. (**c**) True vs. predicted values for node 8 in PEMS-BAY dataset. (**d**) True vs. predicted values for node 66 in PEMS-BAY dataset

0 during data preprocessing. Nonetheless, our model does not force-fit to noisy data and can predict traffic flows accurately.

3.3.6.4 Ablation Study

To assess the efficiency of different modules, a comparison of the following variants was conducted using the METR–LA dataset.

- NNCE. The node correlation estimator module is removed, and the feedforward network is used instead.
- NTRL. The TRL module is removed, using only a single TCN with ELU activation.
- NGC. The gated graph convolutional fusion module is removed, and only adaptive diffusion graph convolution is used.

Table 3.5 presents the average results of the multi-step ablation experiments, where MSGSGCN has the lowest errors in 15 min, 30 min, and 60 min time step predictions, indicating that the addition of the NCE module enables the model to

Table 3.5 Comparison of ablation experiments

Ablation	15 min			30 min			60 min		
	MAE	RMSE	MHD	MAE	RMSE	MHD	MAE	RMSE	MHD
NNCE	2.604	4.897	0.1662	2.867	5.625	0.1699	3.188	6.460	0.1743
NTRL	2.569	4.770	0.1657	2.810	5.423	0.1692	3.111	6.209	0.1735
NGC	2.565	4.771	0.1656	2.813	5.428	0.1692	3.122	6.212	0.1736
MSGSGCN	2.558	4.739	0.1655	2.797	5.392	0.1690	3.093	6.155	0.1732

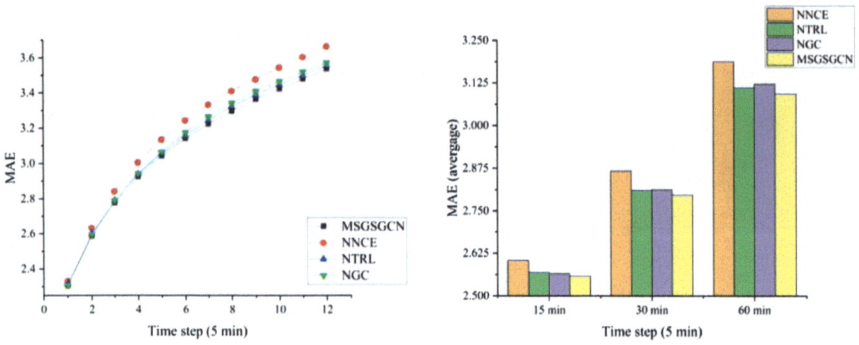

(a) (b)

Fig. 3.15 Comparison with baseline models in terms of MAE. (**a**) METR-LA. (**b**) PEMS-BAY

Table 3.6 Comparison of ablation model complexity

Ablation	Total params (units)	FLOPs (M)	Training time (s/epoch)	Inference time (s)
NNCE	299,620	18,143.29	63.4374	2.3665
NTRL	279,024	15,338.2	58.5805	2.2593
NGC	305,968	13,927.34	40.2114	1.7264
MSGSGCN	312,304	18,159.91	64.8359	2.4975

better learn the dynamic correlations among nodes and improve the accuracy. Removing the TRL module significantly reduces the model's ability to learn long-term traffic data, while removing the gated graph convolutional fusion module leads to the inclusion of too much useless information during the computation process, making it difficult for the model to learn useful information and hence decreasing the prediction accuracy. Furthermore, the MHD evaluation results show that the MSGSGCN maintains good predictive performance across all forecast time steps, indicating that the model possesses a robust noise resistance capability. To illustrate the accuracy improvement more clearly, Fig. 3.15 shows the multi-step error value change.

The significance of removing the contrastive complexity of models is that it helps us understand the contributions and roles of different components of models, optimize model performance, and provide more effective guidance for model design and application. To explore the impact of the improved module on model complexity, Table 3.6 compares the total number of model parameters, model computational power (FLOPs), training time per epoch, and inference time of the ablation model.

It can be seen in Table 3.6 that the removal of the ablation module decreases the complexity and computation time of the model decrease. Among them, the removal of the gated graph convolutional fusion module greatly reduces the computational complexity of the model, which is mainly because this module cycles in the spatiotemporal module. It is worth noting that the NCE module has the least impact on the computational complexity of the model, and its corresponding ablation model also has a good prediction accuracy, indicating that adding this module can effectively improve the prediction accuracy of the model.

3.3.6.5 Parameter Sensitivity

The number of heads in the multi-head self-attention mechanism is also an important parameter, which controls how many heads the input feature map is divided into and computed in parallel. The selection of the number of heads can affect the effect of multi-head self-attention. Experiments were carried out on the datasets METR–LA and PEMS–BAY with the number of heads being 1, 2, 4, 8, and 16, and the results are shown in Fig. 3.16.

It can be obtained from Fig. 3.16 that both too many heads and too few heads have a great influence on the experimental results. When the number of heads is too large, too many heads will lead to low computational efficiency and prone to overfitting. When the number of heads is too small, too few heads will lead to insufficient representation power and prone to underfitting. Therefore, after comprehensively comparing the experimental results of the two datasets, the number of heads was set to 4 in both the METR–LA dataset and the PEMS–BAY dataset in this study.

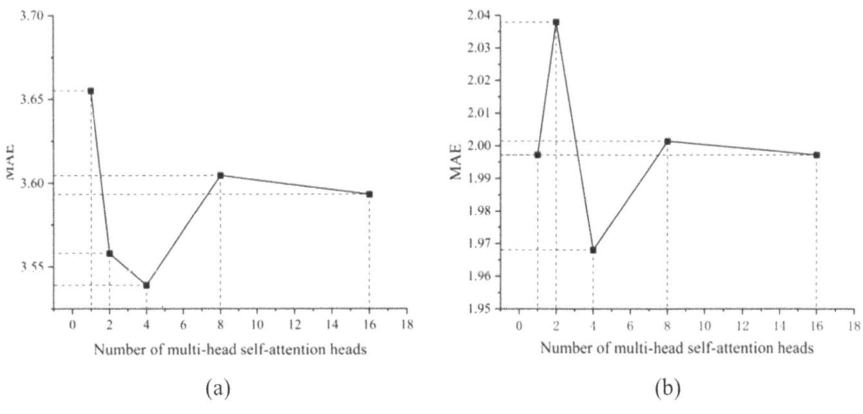

Fig. 3.16 Comparison with baseline models in terms of MAE. (**a**) METR-LA. (**b**) PEMS-BAY

3.3.6.6 Discussion

Compared to the proposed models, the listed baseline models still have shortcomings. STGCN and DCRNN have relatively weak modeling ability for long-term dependencies and suffer from gradient explosion problems. Although the ASTGCN captures the importance of spatiotemporal features using attention mechanisms, it cannot adaptively construct the graph structure. Although the STSGCN constructs a local spatiotemporal graph to learn spatiotemporal features of multiple time steps, its ability to model node dynamics is limited. Although the CCRNN constructs a hierarchical coupling mechanism, it still fails to effectively learn the correlation between nodes. Although the ADN-FA constructs a structure similar to a transformer to learn long-distance temporal dependencies, the fusion of spatiotemporal features is not ideal. In contrast, the MSGSGCN learns the spatiotemporal correlation between nodes and deeply mines the long-term dependent temporal features and deep spatial features, resulting in good prediction performance.

3.4 Dynamic Multiple-Graph Spatial-Temporal Synchronous Aggregation Framework (DMSTSAF)

The overall architecture of our model is presented in Fig. 3.17. We design FAM to take into account the influence of external factors on traffic prediction. Specifically, traffic features and external factors are first concatenated, then fused through spatial and temporal attention mechanisms, and finally mapped to a high-dimensional space and used as inputs to the model. FAM introduces external factors as part of inputs to the model and enhances the representation of traffic features, which addresses the first limitation mentioned in "Introduction." Then, we stack several dynamic multiple-graph spatial-temporal synchronous aggregation layers (DMSTSAL) to model spatial-temporal dependencies. In each layer, we construct two kinds of dynamic spatial-temporal synchronous graphs in parallel: trend spatial-temporal synchronous graph and pattern spatial-temporal synchronous graph. We then deploy a spatial-temporal synchronous aggregation module (STSAM) for each spatial-temporal synchronous graph. The outputs of the two types of modules are fused through a gating mechanism. The design of dynamic multiple graphs instead of a single graph empowers the model to learn spatial-temporal correlations comprehensively, which overcomes the second limitation mentioned in "Introduction." In addition, dilated gated convolution module (DGCM) is also designed to extract long-term dependencies in each layer, and its output is integrated with the outputs of both types of STSAMs. We further design the output module with concatenation and fully connected layers to transform the outputs of DMSTSALs into predictions.

In the next subsections, we describe in detail the various components of DMSTSAF, including FAW, graph construction, STSAM, DTCM, DMSTSL, and output module.

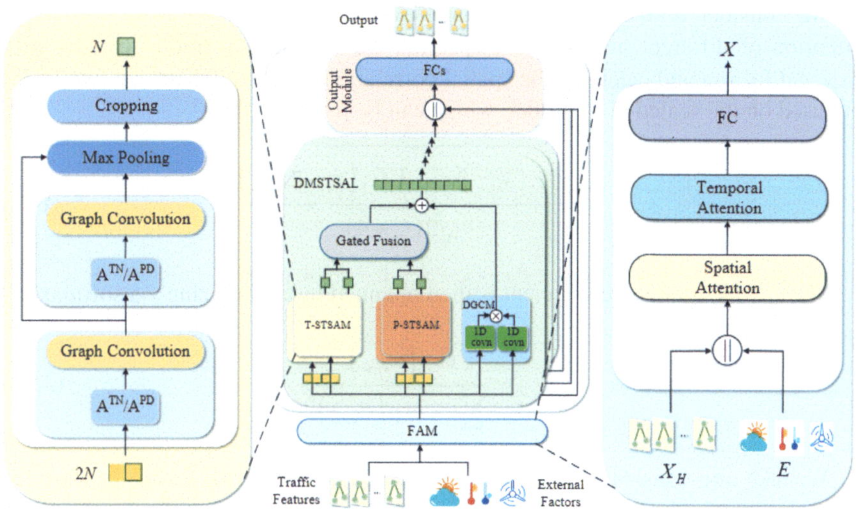

Fig. 3.17 Detailed framework of DMSTSAF, FAM, and STSAM. A DMSTSAF contains an FAW, four DMSTSALs, and an output module. Traffic features and external factors are first concatenated, then spatial attention and temporal attention are applied to enhance representation, and then high-dimensional hidden features are obtained via a fully connected layer. Independent T-STSAMs and parallel P-STSAMs in each DMSTSAL are designed to model spatial-temporal dependencies from multiple perspectives. In STSAMs, stacked graph convolutions followed by the max pooling and the cropping operation are incorporated with spatial-temporal synchronous graphs to extract spatial-temporal features synchronously. Gated 1D convolutions with shared parameters are utilized to learn long-term correlations in DGCM. Finally, the output module yields the predictions of the model

3.4.1 Feature Augmentation Module

Traffic features are affected by external factors such as weather, temperature, wind, etc. In order to integrate external factors with traffic features for accurate traffic prediction, we design FAW, whose structure is shown in Fig. 3.17. The comfort of the temperature is closely related to people's willingness to travel, and the traffic flow in the road network shows varying intensities under different weather conditions (e.g., sunny, rainy). In addition, wind also affects people's travel plans. Therefore, we take account of the effects of four external factors that have the most impact on traffic, namely maximum temperature, minimum temperature, weather, and wind. To better illustrate the point in spatial and temporal attention, we define a fully connected layer as:

$$f(x) = \text{ReLU}(xW + b), \tag{3.32}$$

where W and b are learnable parameters, and ReLU is the activation function.

We consider both traffic features and external factors to measure the spatial attention of different nodes. To be specific, we concatenate traffic features with external factors and calculate the spatial correlation coefficient between two nodes at based on the scaled dot-product, which can be formulated as:

$$\alpha_{v_i,v}^{t_j} = \frac{f_{s,1}\left(x_{v_i,t_j} \| e_{v_i,t_j}\right) f_{s,2}\left(x_{v,t_j} \| e_{v,t_j}\right)}{\sqrt{2(C+4)}}. \tag{3.33}$$

The spatial attention coefficient is then obtained by normalizing via softmax:

$$s_{v_i,v}^{t_j} = \frac{\exp\left(\alpha_{v_i,v}^{t_j}\right)}{\sum_{v\in V}\exp\left(\alpha_{v_i,v}^{t_j}\right)}. \tag{3.34}$$

We take into consideration both traffic features and external factors to learn the temporal attention of various time steps. Specifically, we first concatenate traffic features with external factors, and then employ the scaled dot-product to compute the temporal correlation coefficient. After that, the temporal attention coefficient is obtained by softmax, which can be described as:

$$\beta_{t_j,t}^{v_i} = \frac{f_{u,1}\left(x_{v_i,t_j} \| e_{v_i,t_j}\right) f_{u,2}\left(x_{v_i,t} \| e_{v_i,t}\right)}{\sqrt{2(C+4)}},$$
$$u_{t_j,t}^{v_i} = \frac{\exp\left(\beta_{t_j,t}^{v_i}\right)}{\sum_{t\in T}\exp\left(\beta_{t_j,t}^{v_i}\right)}. \tag{3.35}$$

After obtaining the spatial attention matrix S and the temporal attention matrix U, traffic features and external factors are fused and mapped to the high-dimensional space, and the mathematical expression can be defined as:

$$X = f_x\left(\left(U\cdot\left(S\cdot\left(X_H \| E\right)\right)^T\right)^T\right). \tag{3.36}$$

3.4.2 *Dynamic Spatial-Temporal Synchronous Graph Construction*

Traffic features are related in multiple ways over time and space. On the one hand, nodes in different locations have adjacency and distance relationships. On the other hand, different traffic features may have the same temporal pattern or trend. A single spatial or temporal graph can only focus on one aspect of spatial-temporal dependencies while ignoring others. Aiming to model spatial-temporal correlations from multiple perspectives, we first propose two spatial graphs, i.e., the neighbor graph and the distance graph, as well as two temporal graphs, i.e., the trend graph and the pattern graph, and then design two dynamic spatial-temporal synchronous graphs based on them, i.e., the trend spatial-temporal synchronous graph and the pattern spatial-temporal synchronous graph.

Neighbor graph, which denotes the spatial adjacency of nodes, and the elements of its adjacency matrix are defined as:

$$A_{ij}^{Ne} = \begin{cases} 1, & \text{if } v_i \text{ connects to } v_j, \\ 0, & \text{else.} \end{cases} \tag{3.37}$$

Distance graph represents the distance relationship between nodes, the elements are formulated as:

$$A_{ij}^{Ne} = \begin{cases} 1, & \text{if } v_i \text{ connects to } v_j, \\ 0, & \text{else.} \end{cases} \tag{3.38}$$

Trend graph. In order to capture the trend similarity between traffic features, we propose the trend graph based on dynamic time warping (DTW) algorithm. Given two sequences P and Q, the DTW distance between P and Q can be obtained by the iteration of the following equations:

$$\boldsymbol{M}_D(i,j) = \boldsymbol{M}(i,j) + \boldsymbol{M}_{\min}(i,j). \tag{3.39}$$

The adjacency matrix can be formulated as:

$$A_{ij}^{Tr} = \begin{cases} 1, & \text{if } M_D(i,j) \le \varepsilon_T, \\ 0, & \text{else,} \end{cases} \tag{3.40}$$

Pattern graph. Some traffic nodes exhibit strong linear correlation in their features because they are located in the same functional area (e.g., residential area, commercial area, etc.), as shown in Fig. 3.18b. For the purpose of extracting pattern similarity, we design the pattern graph by Pearson correlation coefficient. The Pearson correlation coefficient between the two nodes can be defined as:

$$\rho_{ij} = \frac{\sum_{i=1}^{T}\left(x_i' - \bar{x}_i\right)\left(x_j' - \bar{x}_j\right)}{\sqrt{\sum_{i=1}^{T}\left(x_i' - \bar{x}_j\right)^2}\sqrt{\sum_{i=1}^{T}\left(x_j' - \bar{x}_j\right)^2}}. \tag{3.41}$$

Dynamic spatial-temporal synchronous graphs. The spatial and temporal correlations of traffic features exist simultaneously and to model the intricate spatial-temporal dependencies synchronously from multiple perspectives, we propose the trend spatial-temporal synchronous graphs GTN and the pattern spatial-temporal synchronous graphs GPD inspired by STSGCN (Song et al., 2020) and STFGNN

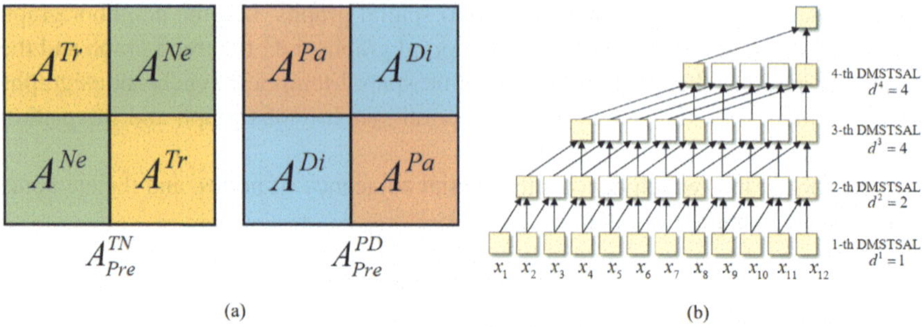

Fig. 3.18 The structures of two predefined adjacency matrices and the dilated step sizes in four layers. (**a**) The structures of the two predefined adjacency matrices. (**b**) The dilated step sizes in four layers

(Li & Zhu, 2021). First, we design two predefined spatial-temporal synchronous graphs, which contain two time steps. The structure of them is illustrated in Fig. 3.18a.

The main diagonals denote that each node has connectivity to nodes with the same trend or pattern at the same time step, while the counter-diagonals indicate that nodes are connected to their neighboring nodes or proximity nodes at the adjacent time step. Then, the two predefined adjacency matrices are multiplied by the learnable parameters of the same shape to obtain the adjacency matrices of the dynamic spatial-temporal synchronous graphs as follows:

$$\begin{aligned} A^{TN} &= W^{TN} \otimes A_{\text{Pre}}^{TN} \in \mathbb{R}^{2N\times 2N}, \\ A^{PD} &= W^{PD} \otimes A_{\text{Pre}}^{PD} \in \mathbb{R}^{2N\times 2N}. \end{aligned} \tag{3.42}$$

3.4.3 Experiments

3.4.3.1 Datasets

Four public datasets are chosen for evaluating the prediction performance of DMSTSAF: PEMS03, PEMS04, PEMS07, and PEMS08. All datasets are generated by Caltrans Performance Measurement System (PeMS). Concretely, traffic features in these datasets are collected by sensors located on California highways at 5-min intervals, which means there are 12 time steps in an hour. Spatial graphs are constructed from the distribution of sensors. The distinctions among the datasets are the geographic locations of sensors and the temporal ranges of data. The detailed information of four datasets can be found in Table 3.7.

Table 3.7 Detailed information of datasets

Datasets	Samples	Number of nodes	Traffic features	Time span
PEMS03	26,208	358	Flow	09/01/2018–11/30/2018
PEMS04	16,992	307	Flow, speed, occupancy	01/01/2018–02/28/2018
PEMS07	28,224	883	Flow, speed, occupancy	05/01/2017–08/31/2017
PEMS08	17,856	170	Flow	07/01/2016–08/31/2016

3.4.3.2 Experiment Settings and Evaluation Metrics

In our experimental implementation, each dataset is partitioned for training, validation, and testing in a ratio of 60%, 20%, and 20%. We employ the traffic flow of 12 historical time steps (1 h) to predict the traffic flow of 12 future time steps. The traffic flow in each dataset is standardized using *Z*-score normalization.

DMSTSAF is performed by Pytorch using a PC with NVIDIA RTX 3080. We chose hyperparameters of our model experimentally to ensure superior performance. The sparsity of the adjacency matrices for the distance graph, the trend graph, and the pattern graph are set as 0.01. Each STSAM consists of two graph convolutions, and the output dimension of the fully connected layer in FAW and the hidden dimensions of graph convolutions in STSAM are set as D ¼ 64. The hidden representations of two fully connected layers to generate predictions are tuned as [128,1], respectively. The optimizer in experiments is set as Adam. The batch sizes for four datasets are tuned as [32,32,8,64], respectively. The initial learning rate is set as 0.003 and scaled down to 0.3 times every 30 epochs to shorten the training time. Enrol up to 200 epochs in each training.

We take three metrics to evaluate the performance of models, which are mean absolute error (MAE), mean absolute percentage error (MAPE), and root mean square error (RMSE). Smaller values indicate better performance for these metrics.

3.4.3.3 Baseline Methods

- FC-LSTM: Long short-term memory network with fully connected layers, a variant of the recurrent neural network, which consists of the forget gate, the input gate, and the output gate.
- GRU: Gate recurrent unit, a variant of LSTM, which simplifies the three gates in LSTM to two, i.e., the reset gate and the update gate.
- T-GCN: Temporal graph convolutional network, which incorporates GCN with GRU to extract spatial-temporal features.
- DCRNN: Diffusion convolution recurrent neural network, in which random walks of a graph are utilized to model spatial dependence, GRU is employed to learn temporal correlation.
- STGCN: Spatiotemporal graph convolution network, which takes advantage of GCN to extract spatial features, and makes use of 1D convolution rather than recurrent neural network to model temporal dependency.

- STSGCN: Spatial-temporal graph convolutional network, which constructs localized spatial-temporal graphs based on spatial and temporal adjacencies, and designs spatial-temporal synchronous graph convolutional modules to learn spatial temporal correlations synchronously.
- STFGNN: Spatial-temporal fusion graph neural network, an improved study based on STSGCN, which designs temporal graphs with dynamic time warping algorithm and proposes spatial-temporal fusion graphs to model localized spatial temporal dependencies, and then designs gated CNN to extract long-range dependencies.

3.4.3.4 Experimental Results

The experiments of traffic flow prediction are performed on four datasets, and a comparison of the results for all models is shown in Table 3.8. DMSTSAF obtains the smallest metrics on all four datasets, demonstrating the consistent superiority of DMSTSAF over the baselines. Compared to the state-of-the-art baseline STFGNN, DMSTSAF improves 4.27%, 3.72%, 8.54%, and 7.89% in terms of MAE on PEMS03, PEMS04, PEMS07, and PEMS08, respectively, while the improvements of MAPE are 6.63%, 7.49%, 7.95%, and 7.85%. In addition, our model also achieves 3.68%, 4.23%, 8.38%, and 7.65% improvements in terms of RMSE.

Moreover, for the purpose of evaluating the ability of models on multi-step prediction, we conduct prediction experiments for 12 future time steps on PEMS03. In comparison to STFGNN, our model shows 1.11–8.89% improvement in terms of MAE, 1.81–13.19% improvement for MAPE, and 1.38–7.48% improvement for RMSE. To achieve a more intuitive comparison, we illustrate the results of each

Table 3.8 Results of DMSTSAF and baselines on four datasets

Datasets	Metric	FC-LSTM	GRU	T-GCN	DCRNN	STGCN	STSGCN	STFGNN	DMSTSAF
PEMS03	MAE	27.35	27.19	20.83	21.07	20.29	18.30	17.33	16.59
	MAPE (%)	25.16	24.92	21.58	20.43	18.98	17.58	16.90	15.78
	RMSE	42.32	42.24	31.18	33.23	33.08	30.20	29.04	27.97
PEMS04	MAE	34.33	34.13	26.02	27.57	25.37	22.38	20.14	19.39
	MAPE (%)	21.72	21.53	17.08	18.29	15.50	15.03	13.76	12.73
	RMSE	50.01	49.98	38.19	42.07	39.11	35.33	32.62	31.24
PEMS07	MAE	38.33	37.79	30.37	31.29	30.91	25.15	23.76	21.73
	MAPE (%)	16.71	16.83	13.83	15.09	14.49	10.74	9.96	9.17
	RMSE	57.56	56.72	43.39	47.21	47.12	40.79	38.48	35.25
PEMS08	MAE	28.90	28.12	21.37	21.21	21.39	17.72	16.99	15.65
	MAPE (%)	17.77	16.92	13.66	13.08	13.13	11.61	10.96	10.10
	RMSE	41.83	41.85	30.69	31.43	31.42	27.21	26.80	24.75

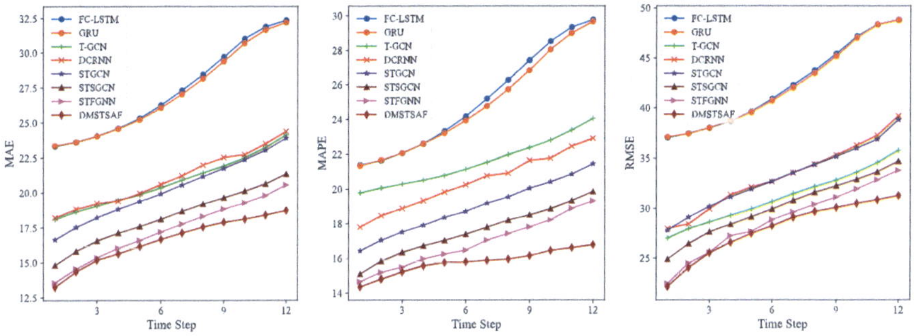

Fig. 3.19 Results of multi-step prediction on PEMS03

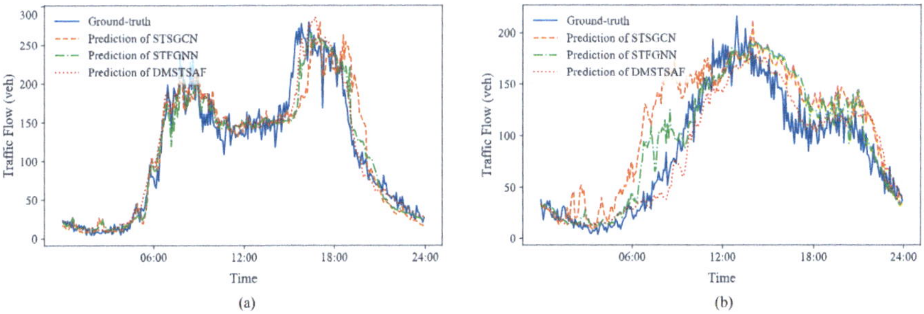

Fig. 3.20 Prediction results of STSGCN, STFGNN, DMSTSAF. (**a**) Prediction results on a sunny day. (**b**) Prediction results on a rainy day

model with a line plot, as shown in Fig. 3.19, which verifies that DMSTSAF overwhelmingly outperforms all the baselines.

To further demonstrate the prediction performances of DMSTSAF and the best baselines, we visualize the ground-truth and predictions of STSGCN, STFGNN, and DMSTSAF on PEMS03 test set for both sunny and rainy days, as presented in Fig. 3.20. It can be seen that compared to the best baselines, our proposed DMSTSAF can fit the ground-truth more accurately in different weather conditions.

3.4.3.5 Ablation Study

To evaluate the validity of different modules in DMSTSAF, ablation studies are implemented on PEMS04 and PEMS08. We design six variants, a short illustration is introduced as follows:

- -FAW, which uses only traffic features as the input to the model and utilizes an FC rather than a feature augmentation module to transform the input from C dimensions to D dimensions.

- -GTN, which removes trend spatial-temporal synchronous graphs GTN and the corresponding T-STSAM, and retains only pattern spatial-temporal synchronous graphs GPD as well as P-STSAM.
- -GPD, which is the opposite of -GTN, removing GPD and P-STSAM, while retaining GTN as well as T-STSAM.
- -Dilation, the step sizes of 1D convolutions in DGCM and the distance between two time steps to construct dynamic spatial-temporal synchronous graphs are fixed to 1.
- -Concatenation, which removes the concatenation on the outputs of four DMSTSALs and treats the output of the 4-th DMSTSAL as the input to the output module.
- -DGCM, which removes dilated gated convolution modules from four DMSTSALs.

Table 3.9 demonstrates the results of ablation studies, which illustrate that DMSTSAF outperforms variants on PEMS04 as well as PEMS08. Compared to the three indicators of -FAW, our model achieves 1.92%, 2.60% as well as 1.67% improvement on PEMS04. Furthermore, it improves by 3.04%, 2.79%, and 1.39% on PEMS08. These improvements indicate the effectiveness of fusing traffic features and external factors in learning spatial-temporal correlations. In contrast with -GTN, DMSTSAF achieves 1.02%, 1.53%, 1.01% improvements on PEMS04, as well as 1.39%, 2.23%, 0.88% improvements on PEMS08. In comparison to -GPD, improvements of 0.6%, 0.94%, 0.54% on PEMS04 are obtained by our model, while improvements of 0.70%, 1.56%, 0.52% are achieved on PEMS08. These achievements verify the validity of dynamic multiple-graph in modeling spatial-temporal dependencies. Compared to -DGCM, our model achieves 1.87%, 2.68%, as well as 2.74% improvement on PEMS04. Furthermore, it improves 2.49%, 4.54%, and 1.67% on PEMS08. These improvements illustrate the positive role of DGCM in capturing long-term correlations. In addition, there are also different levels of improvement in our model compared to other variants, which demonstrates the efficiencies of the corresponding components.

Table 3.9 Results of ablation studies

		PEMS04			PEMS08	
Model and variants	MAE	MAPE (%)	RMSE	MAE	MAPE (%)	RMSE
DMSTSAF	19.39	12.73	31.24	15.65	10.10	24.75
-FAW	19.77	13.07	31.77	16.14	10.39	25.10
-*ATN*	19.59	12.93	31.56	15.87	10.33	24.97
-*APD*	19.46	12.85	31.41	15.76	10.26	24.88
-Dilation	19.62	13.05	31.61	15.95	10.37	25.05
-Concatenation	20.12	13.35	32.52	17.01	10.89	26.67
-DGCM	19.76	13.08	32.12	16.05	10.58	25.17

3.4.3.6 Parameters Study

In order to further validate our model, parameters study are implemented on PEMS04 and PEMS08. The number of graph convolutions and the hidden dimensions D in STSAM are under consideration, Fig. 3.21 presents the results of experiments. Our model achieves minimal errors as each STSAM consists of two graph convolutions. As for the hidden dimensions, it can be found in Fig. 3.21 that our model obtains optimal metrics when D equals to 64.

3.4.3.7 Discussion

By means of the results, it can be inferred that our proposed model outperforms the seven baselines. FC-LSTM and GRU are not satisfactory since they only take into account temporal dependency and ignore spatial relationships. T-GCN, DCRNN,

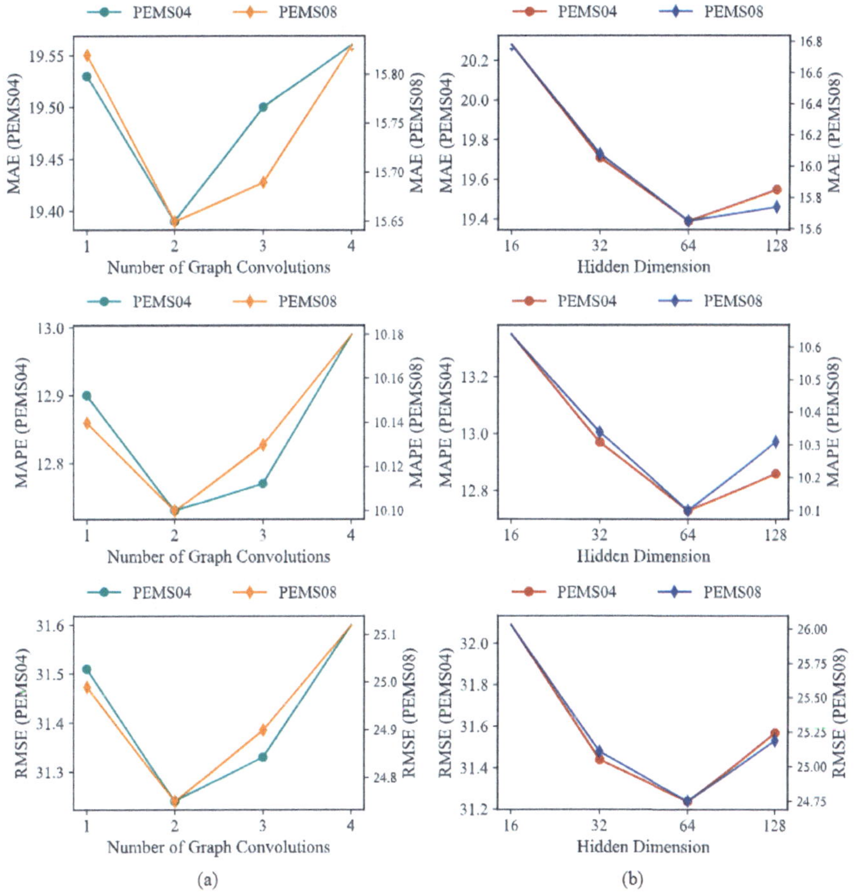

Fig. 3.21 Experimental results of parameters on PEMS04 and PEMS08. (a) Results for different numbers of graph convolutions. (b) Results for different hidden dimensions

and STGCN utilize graph convolution network to model spatial dependence and further employ temporal components to extract temporal features, leading to improved performance than models only applicable for time sequences. However, their separate modules for spatial and temporal modeling limit the efficiency of feature extraction.

STSGCN designs synchronous graphs to characterize spatial-temporal correlations simultaneously. Further, STFGNN proposes temporal graphs generated by a dynamic time warping algorithm, and then designs spatial-temporal fusion graphs. Although they both achieve synchronous learning compared to previous work, their performances are still insufficient. First, neither of them takes account of the impact of external factors, which weakens the ability of models to extract comprehensive dependencies. Second, their single spatial-temporal graph structures result in overlooking certain potential correlations. Our model integrates maximum temperature, minimum temperature, weather, as well as wind with traffic features by spatial and temporal attention, and then feeds the fused features into subsequent modules. Therefore, the influence of external factors is taken into consideration in spatial-temporal modeling. In addition, we introduce four graphs that lead to two dynamic spatial-temporal synchronous graphs, bringing about multi-perspective spatial-temporal simultaneous modeling of hidden features, enhancing the proposed model's performance for learning spatial and temporal dependencies.

Ablation studies illustrate the effectiveness of various components in our model. FAW is effective for feature extraction because traffic features vary significantly with different external factors. Both GTN and GPD contribute noticeably to the improved performance of DMSTSAF due to their capacity for learning spatial-temporal correlations from different perspectives. Other components are also beneficial to strengthening the model's performance. Specifically, dilated step sizes expand the receptive fields of stacked layers, the concatenate operation on outputs of four layers retains more useful hidden information, and DGCM captures long-term temporal dependence.

Furthermore, the excellent performance of DMSTSAF benefits from the appropriate number of graph convolutions and the suitable hidden dimensions. When each STSAM comprises only one graph convolution, this leads to the inefficient representation of hidden dimensions, while three or four graph convolutions consisting of one STSAM result in over-smoothing. If we tuned hidden dimensions in STSAM to 16 or 32, the inability to obtain efficient deep representations restricts the capacity to model dependencies, but tuning hidden dimensions to 128 results in overfitting.

Although diverse experiments confirm the effectiveness of our model, there are still some shortcomings. First, the proposed method cannot efficiently deal with the missing data in datasets. Second, the external factors only include meteorological conditions, and further extensions are yet to be made. In future work, we plan to introduce processing algorithms for missing data and incorporate additional external factors to improve our model further.

3.5 Conclusion

In conclusion, this chapter emphasizes the critical role of dynamic and multi-graph approaches in connected traffic flow prediction. By utilizing multiple graph structures to aggregate and analyze spatiotemporal relationships, these methods effectively capture the intricacies of traffic behavior under varying conditions. The examination of models such as the multi-sequential temporal convolution gated graph neural network and the dynamic multi-graph synchronous aggregation framework reveals their potential to enhance prediction capabilities in complex urban environments. This chapter ultimately underscores the importance of adapting to dynamic traffic patterns, paving the way for more accurate and responsive traffic prediction systems.

References

1. S. Lu, Q. Zhang, G. Chen, D. Seng, A combined method for short-term traffic flow prediction based on recurrent neural network. Alex. Eng. J. **60**(1), 87–94 (2021). https://doi.org/10.1016/j.aej.2020.06.008
2. P. Xie, T. Li, J. Liu, S. Du, X. Yang, J. Zhang, Urban flow prediction from spatiotemporal data using machine learning: A survey. Inf. Fusion **59**, 1–12 (2020). https://doi.org/10.1016/j.inffus.2020.01.002
3. S.A. Seshia, S. Jha, T. Dreossi, Semantic adversarial deep learning. IEEE Des. Test **37**(2), 8–18 (2020). https://doi.org/10.1109/mdat.2020.2968274
4. Y. Tian, K. Zhang, J. Li, X. Lin, B. Yang, LSTM-based traffic flow prediction with missing data. Neurocomputing **318**, 297–305 (2018). https://doi.org/10.1016/j.neucom.2018.08.067
5. C. Ma, Y. Zhao, G. Dai, X. Xu, S.-C. Wong, A novel STFSA-CNN-GRU hybrid model for short-term traffic speed prediction. IEEE Trans. Intell. Transp. Syst. **24**(4), 3728–3737 (2023). https://doi.org/10.1109/tits.2021.3117835
6. Y. Bao, J. Liu, Q. Shen, Y. Cao, W. Ding, Q. Shi, PKET-GCN: Prior knowledge enhanced time-varying graph convolution network for traffic flow prediction. Inf. Sci. (NY) **634**, 359–381 (2023). https://doi.org/10.1016/j.ins.2023.03.093

Chapter 4
Hybrid Models Leveraging Local and Global Spatial Correlation for Traffic Prediction

4.1 Global Spatial Correlation

Traditional GCN propagation relies on the actual topological structure of the road network and does not consider the varying spatial impact between adjacent road segments. In cases where the correlation between some non-adjacent road segments is greater than the correlation between certain adjacent road segments, inputting the original adjacency matrix may lead to the loss of crucial spatial information for certain important road segments. This could result in suboptimal predictive outcomes. To better explore the traffic characteristics of urban road networks, this paper proposes a speed prediction model named G-GCGRU based on global spatio-temporal features. The model utilizes global graph convolutional networks for spatial feature analysis, constructing an internally spatial relationship with directed and weighted road networks. This approach addresses the issue of missing information on non-adjacent highly correlated road segments, allowing for a comprehensive exploration of spatial features in traffic flow. Simultaneously, the model employs GRU to analyze the temporal dependence of the data, uncovering its temporal characteristics.

4.2 A Hybrid Model Integrating Local and Global Spatial Correlation (T-LGGCN)

4.2.1 *Spatial Correlation*

GCN is a popular neural network for processing spatial correlation. The topology of the road network is represented as an undirected graph $\boldsymbol{G} = (\boldsymbol{V}, \boldsymbol{E})$, where $\boldsymbol{V}$ is the set of nodes representing the sensors, and $\boldsymbol{E}$ is the set of edges, denoting the

Q. Shi et al., *Connected Vehicles Traffic Prediction*, Wireless Networks, https://doi.org/10.1007/978-3-031-84548-2_4

adjacency of the sensors. We can convert the adjacency of the graph $\boldsymbol{G}$ into the adjacent matrix $\boldsymbol{A}$ as follows:

$$a_{ij} = \begin{cases} e^{-\frac{\text{dist}(i,\,j)^2}{\sigma^2}} & \text{dist}(i,j) < \lambda \\ 0 & \text{dist}(i,j) \geq \lambda \end{cases} \tag{4.1}$$

where a_{ij} represents the adjacent weight between sensor i and sensor j, dist(i, j) is the distance between sensor i and sensor j, σ^2 is the variance, and λ is the threshold.

Then the propagation rule of GCN is as follows:

$$\boldsymbol{H}^{(l+1)} = \sigma\left(\hat{\boldsymbol{A}}\boldsymbol{H}^{(l)}\boldsymbol{W}^{(l)}\right) \tag{4.2}$$

where $\hat{\boldsymbol{A}} = \tilde{\boldsymbol{D}}^{-1/2}\tilde{\boldsymbol{A}}\tilde{\boldsymbol{D}}^{-1/2}$, $\tilde{\boldsymbol{A}} = \boldsymbol{I} + \boldsymbol{A}$, $\boldsymbol{I} \in R^{N \times N}$ is the identity matrix, $\tilde{D}_{ii} = \sum_j \tilde{A}_{ij}$, $\boldsymbol{H}^{(l)}$ denotes the output of the lth layer, $\boldsymbol{W}^{(l)}$ is the weight parameter matrix of the lth layer.

Although the adjacency matrix can show the intuitive sensors' adjacency, it does not express the internal spatial influence. Therefore, we analyze the spatial correlation from two perspectives, global spatial correlation and local spatial correlation, respectively.

4.2.1.1 Local Spatial Correlation Component

It can be found that the more adjacent nodes, the smaller the weight of its node will be considered in the feature aggregation process during the convolution operation of Eq. (4.3). Therefore, in order to mine the spatial characteristics of the nodes themselves, we use the approximate personalized propagation of neural predictions (APPNP) model to mine the local spatial correlation of traffic flow. APPNP utilizes PageRank for node feature propagation, using PageRank to encode features for each root node and increase the chance of transmission back to the root node. In this way, the model can balance the need of retaining local features and mining neighborhood features. The model calculation rules are as follows:

$$\boldsymbol{Z}^{(0)} = \boldsymbol{H} = f_\theta(\boldsymbol{X}) \tag{4.3}$$

$$\boldsymbol{Z}^{(k+1)} = (1-\alpha)\hat{\boldsymbol{A}}\boldsymbol{Z}^{(k)} + \alpha\boldsymbol{H} \tag{4.4}$$

where $\boldsymbol{X}$ represents the input of the nodes, f_θ denotes a neural network. We use f_θ to extract each sensor's self-features, α to represent the percentage of self-features. As shown in Fig. 4.1, we mine the spatial correlation of the nodes themselves and their first-order neighbors, so we set k to 1. The local spatial correlation mining formulas are as follows:

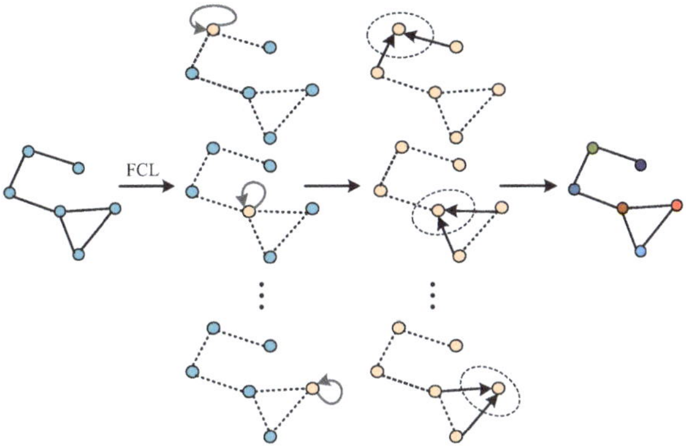

Fig. 4.1 Local spatial process

$$Z^{(0)} = W_L^{(1)} X + b_L^{(1)} \tag{4.5}$$

$$\mathrm{LGCN}(X, A) = \sigma\left(\left(1-\alpha\right)\hat{A}Z^{(0)} + \alpha Z^{(0)}\right) \tag{4.6}$$

We use the FCL to extract node features. $W_L^{(1)}$ represents the weight matrix of the FCL, and $b_L^{(1)}$ is the bias matrix. LGCN(·) denotes the output of local spatial correlation.

4.2.1.2 Global Spatial Correlation Component

Spatial correlation does not only exist between neighboring sensors. As far as the whole road network is concerned, spatial correlation exists between sensors separated by long distances. So, we implicitly express the global spatial correlation of the road network. For sensors data, we use the Pearson correlation coefficient method to analyze the correlation between sensors in the studied network. We set a correlation threshold k to select high correlation sensors. If the correlation is greater than k, the correlation value is kept; otherwise, it is set to 0. In this way, we construct the correlation matrix C. And then, we use it to aggregate the highly correlated sensors' features through the GCN convolution method.

The correlation between two sensors is analyzed through the Pearson correlation coefficient method, which is calculated as follows:

$$C_{ij} = \frac{\sum_{t=1}^{T}\left(x_t^i - \bar{X}_i\right)\left(x_t^j - \bar{X}_j\right)}{\sqrt{\sum_{t=1}^{T}\left(x_t^i - \bar{X}_i\right)^2}\sqrt{\sum_{t-1}^{T}\left(x_t^j - \bar{X}_j\right)^2}} \tag{4.7}$$

where X_i represents the feature of traffic flow of sensor i, $\bar{X}_i$ is the mean value of X_i.

By the convolution of the correlation matrix and the feature matrix, the high correlation node features can be aggregated, which can deeply mine global spatial

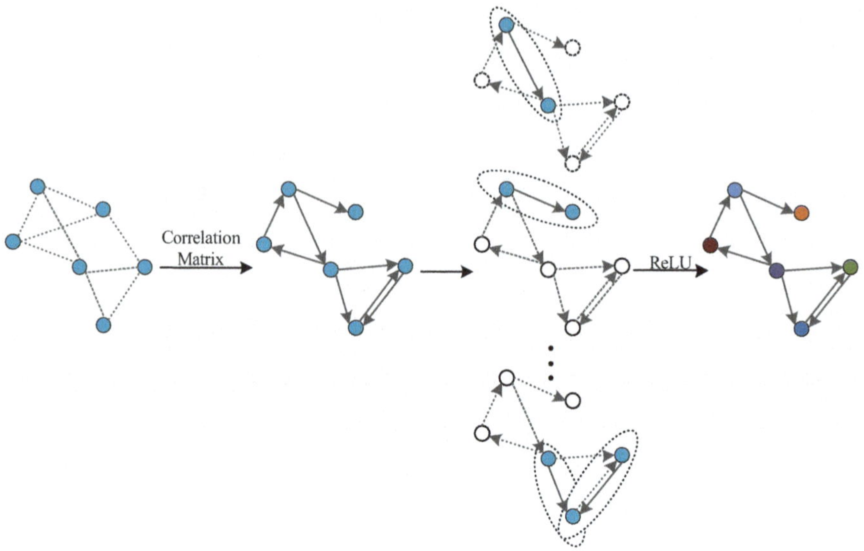

Fig. 4.2 Global spatial process

correlations, as shown in Fig. 4.2. Therefore, the calculation rule of the global graph convolutional network based on the correlation matrix is updated as follows:

$$\mathrm{GGCN}(\boldsymbol{X},\boldsymbol{C})=\sigma\left(\boldsymbol{C}\boldsymbol{X}\boldsymbol{W}_G^{(1)}\right) \tag{4.8}$$

where $\boldsymbol{W}_G^{(1)}$ represents the weight matrix of the global graph convolutional network, GGCN(·) is the output of the global spatial correlation.

4.2.2 *Temporal Correlation*

GRU is a mainstream neural network that addresses time series prediction problems. It can avoid gradient explosion and disappearance of RNN. GRU contains three parts: the input layer, the hidden layer, and the output layer. The core algorithm lies in the computation process in the unit block of the hidden layer, as shown in Fig. 4.3. The local and global spatial correlation output is input into GRU separately. In this way, GRU saves the traffic information of the previous moment and simultaneously combines the traffic context of the current moment, thus, achieves the temporal correlation.

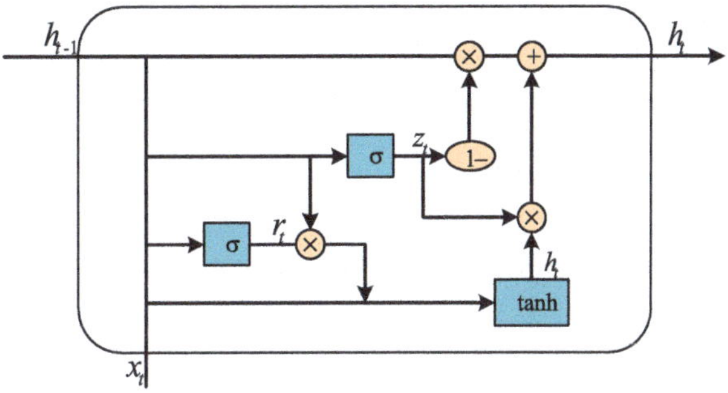

Fig. 4.3 Gated recurrent unit network

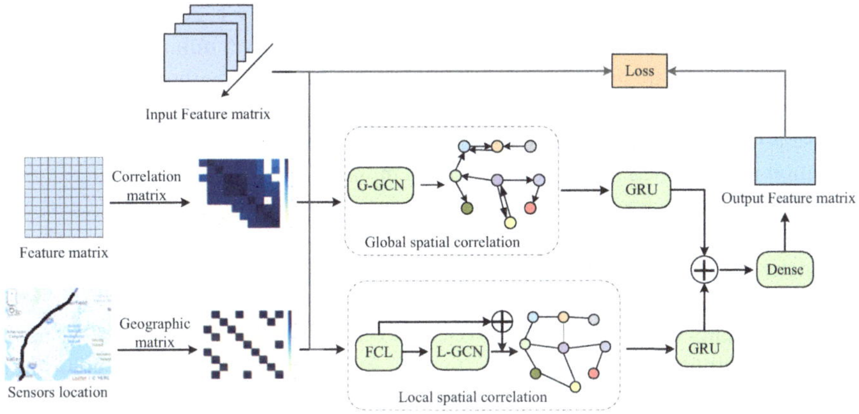

Fig. 4.4 Structure of ST-LGGCN

4.2.3 *T-LGGCN Model*

To address the spatial-temporal correlation of traffic flow, especially the spatial correlation, we respectively construct the global spatial-temporal component and local spatial-temporal component to mine it, as shown in Fig. 4.4.

For the global spatial-temporal component, the global correlation matrix $\boldsymbol{C}$ is first calculated using the full amount of feature data. And then, we feed the input $\boldsymbol{X}$ and $\boldsymbol{C}$ into the global spatial correlation component to obtain GGCN(·), which is put into the GRU to extract the temporal correlation and get the output h_t^g of this component.

Next, the local spatial-temporal component is calculated. We employ the fully connected layer to extract the node features of input $\boldsymbol{X}$ and use the GCN to implement the aggregation and propagation of spatial features. The output LGCN(·) is directly input into GRU to obtain the output h_t^l of this component.

We use Eq. (4.9) to sum the output of the global spatial-temporal component and the local spatial-temporal component and input them into the dense layer to output the prediction results.

$$y_{\text{pre}} = \text{Dense}\left(h_t^l + h_t^g\right) \tag{4.9}$$

4.2.4 Experiments

4.2.4.1 Experiments Settings

Datasets

We evaluated the performance of T-LGGCN through two real highway datasets: the PEMS04 dataset and the PEMS08 dataset. Both datasets are from the Caltrans performance evaluation system. We took the traffic speed data as the traffic flow information. PEMS04 was collected from San Francisco Bay. It contains 307 sensors, and the time is from January 1, 2018, to February 28, 2018. PEMS08 was from San Bernardino. It contains 170 sensors, and the time is from July 1, 2016, to August 31, 2016. The traffic data in both datasets were aggregated every 5 min. There were 288 records per day and no missing data in both datasets. We used the threshold Gaussian kernel to calculate the adjacency matrix of both datasets. During the experiments, we use 80% of the data as the training set and 20% for testing. The normalization operation is performed first when the data is input.

Parameter Settings

The experiments were implemented based on the TensorFlow framework. We used the Adam optimizer to optimize the model. Adam is an optimization algorithm designed to find the global optimal points. It combines the advantages of momentum and RMSProp to minimize the loss function. Adam optimizer is an optimizer that implements Adam algorithm in TensorFlow which has been validated on a large number of neural network experiments. After repeated experiments, when the model performance reached the optimum, the main parameters were set as follows: α is set to 0.8, the GRU dimension is set to 64, the input step is set to 12, the learning rate is set to 0.001, the batch size is set to 32, and the training epoch is set to 100.

Evaluation Metrics

We use the root mean square error (RMSE), mean absolute error (MAE), accuracy, coefficient of determination (R^2) as the evaluation metrics, and the detailed definitions are as follows.

$$\mathrm{RMSE} = \sqrt{\frac{1}{IS}\sum_{i=1}^{I}\sum_{s=1}^{S}\left(y_i^s - y_i^{s'}\right)^2} \tag{4.10}$$

$$\mathrm{MAE} = \frac{1}{IS}\sum_{i=1}^{I}\sum_{s=1}^{S}\left|y_i^s - y_i^{s'}\right| \tag{4.11}$$

$$\text{Accuracy} = 1 - \frac{Y - Y_F'}{Y_F} \tag{4.12}$$

$$R^2 = 1 - \frac{\sum_{i=1}^{I}\sum_{s=1}^{S}\left(y_i^s - y_i^{s'}\right)^2}{\sum_{i=1}^{I}\sum_{s=1}^{S}\left(y_i^s - \overline{y}\right)^2} \tag{4.13}$$

where I represents the input time steps, S is the numbers of sensors, y_i^s and $y_i^{s'}$ denote the real traffic speed data and predicted data, respectively, $\boldsymbol{Y}$ and $\boldsymbol{Y'}$ are the sets of y_i^s and $y_i^{s'}$, $\overline{y}$ is the mean value of $\boldsymbol{Y}$.

4.2.4.2 Baselines

To demonstrate the effectiveness of our model, we select parametric methods (i.e., HA and ARIMA), nonparametric methods (i.e., SVR), and deep learning methods (i.e., GCN, GRU, and T-GCN) for comparison.

1. HA: HA is the simplest parametric method. The average value of historical traffic parameters is used as the model prediction result in our experiments.
2. ARIMA: The basic idea of ARIMA is to treat the time series data as a random series and describe this series approximately with a certain mathematical model. A smoothness test is performed on the data first, and an ARMA model is fitted for prediction. We set the value of the autoregressive coefficient to 1, and the values of the difference order and the moving average to 0.
3. SVR: SVR is one of the common nonparametric prediction methods. The data are mapped to a multidimensional space using a nonlinear function, and then linear regression is performed on them.
4. GCN: GCN is a neural network method for spatial correlation analysis. A one-layer graph convolutional network is used, and the specific calculation process is detailed in Eq. (4.3).
5. GRU: GRU is a variant of RNN. It is usually used to analyze the time series data. The data is directly input into GRU. We set the number of hidden layer neurons to 64.
6. Temporal Graph Convolutional Network(T-GCN): T-GCN is a hybrid neural network. It consists of one-layer GCN and GRU. The feature data is input into GCN for calculation, and the output of the GCN layer is input into GRU to get the prediction results.

4.2.4.3 Experiments Results

Tables 4.1 and 4.2 show the prediction performance of our model and other baseline models on the PEMS04 and PEMS08 datasets for 5 min prediction task. It can be seen that T-LGGCN obtains the best prediction performance under all metrics.

Table 4.1 Prediction results of different models on PEMS04

	PEMS04			
Model	RMSE	MAE	Accuracy	R^2
HA	3.8387	1.7139	0.9402	0.7829
ARIMA	6.7931	4.8239	0.8921	–
SVR	2.5629	1.2432	0.9601	0.9032
GRU	2.5668	1.5293	0.9601	0.9029
GCN	2.9232	1.8483	0.9545	0.8741
T-GCN	2.1204	1.2027	0.9670	0.9337
T-LGGCN	**1.8627**	**1.0088**	**0.9710**	**0.9489**

Table 4.2 Prediction results of different models on PEMS08

	PEMS08			
Model	RMSE	MAE	Accuracy	R^2
HA	3.1179	1.3641	0.9513	0.7982
ARIMA	5.0256	3.5817	0.9201	–
SVR	2.1578	1.0109	0.9663	0.9033
GRU	1.6257	0.8733	0.9746	0.9451
GCN	2.1545	1.2631	0.9663	0.9036
T-GCN	1.9876	1.1737	0.9697	0.9216
T-LGGCN	**1.5287**	**0.8382**	**0.9761**	**0.9515**

We can see that the hybrid neural networks outperform the single neural networks. Compared with GCN and GRU, the RMSE of T-LGGCN on the PEMS04 is improved by 36.2% and 27.4%. This indicates that the hybrid model takes into account both temporal and spatial features, which can more comprehensively explore the traffic flow characteristics.

It is noted from the table that T-LGGCN has a good ability to mine global and local spatial features. Compared with the T-GCN, the RMSE of T-LGGCN decreased by 12.1% and 23.1% on the PEMS04 and PEMS08. T-GCN only aggregates the spatial information of adjacent first-order sensors. T-LGGCN combines the spatial features of the node itself, first-order neighboring sensor spatial information, and long-range highly correlated sensor spatial information, so as to analyze the spatial correlation comprehensively.

To have a deeper understanding of the T-LGGCN performance, we visualize the prediction results of one sensor data on two test sets separately, as shown in Fig. 4.5. We visualize the prediction results for 1 day and 1 week on two datasets. It is observed that our model fits better.

Meanwhile, we compare the prediction results of T-GCN with T-LGGCN, as shown in Fig. 4.6. Obviously, the fit of T-GCN is poorer than that of T-LGGCN. Although the T-GCN prediction results are mostly close to the true values, we can find that the T-GCN does not accurately capture the traffic flow characteristics when the traffic flow state changes drastically. The T-GCN fit curve behaves

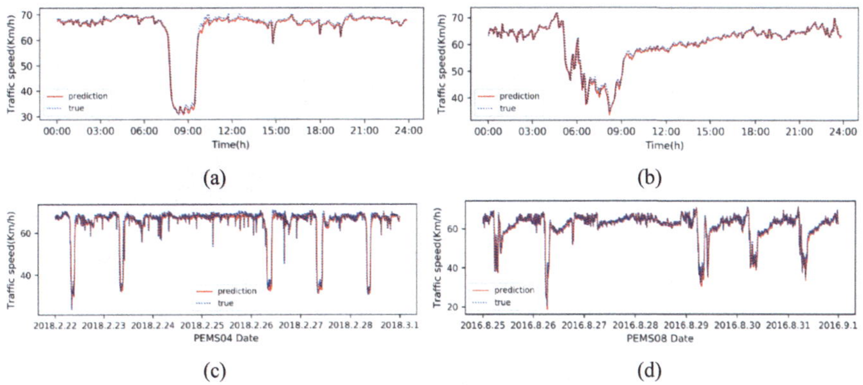

Fig. 4.5 The visualization results in test set. (**a**) T-LGGCN predictions for 1 day on PEMS04. (**b**) T-LGGCN predictions for 1 day on PEMS08. (**c**) T-LGGCN predictions for 1 week on PEMS04. (**d**) T-LGGCN predictions for 1 week on PEMS08

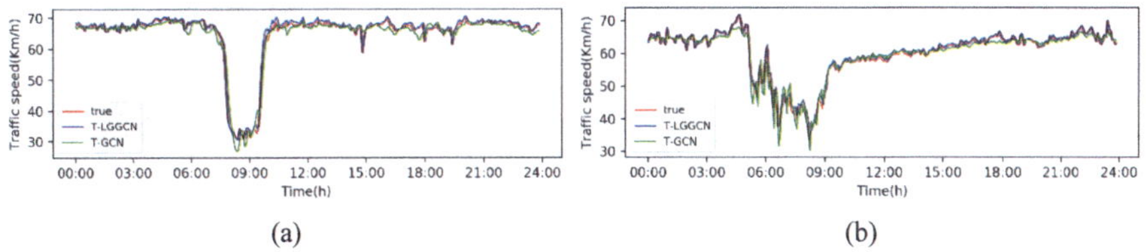

Fig. 4.6 The comparison of results in test set with T-GCN. (**a**) The comparison of results for 1 day on PEMS04. (**b**) The comparison of results for 1 day on PEMS08

relatively smoothly in adjacent times (shown in the orange box). This is because T-GCN uses GCN to perform spatial feature aggregation operations, which may cause an oversmoothing problem, resulting in smoother predicted values at the peak. We improved the GCN spatial aggregation method, and it can be seen that our model effectively avoids this drawback.

4.2.4.4 Experiments Discussion

Spatial Correlation Analysis

T-LGGCN considers the spatial correlation between sensors in the global network. We use Pearson correlation coefficients to analyze the correlation between sensors based on traffic speed. The correlation values of each sensor are shown in Fig. 4.7. Taking the PEMS04 dataset as an example, Fig. 4.7a shows the correlation of the true first-order topological correlation, and Fig. 4.7b shows the sensor correlation used in our model. The darker the color, the higher the correlation. It can be found that the correlation in Fig. 4.7a is significantly weaker than that in

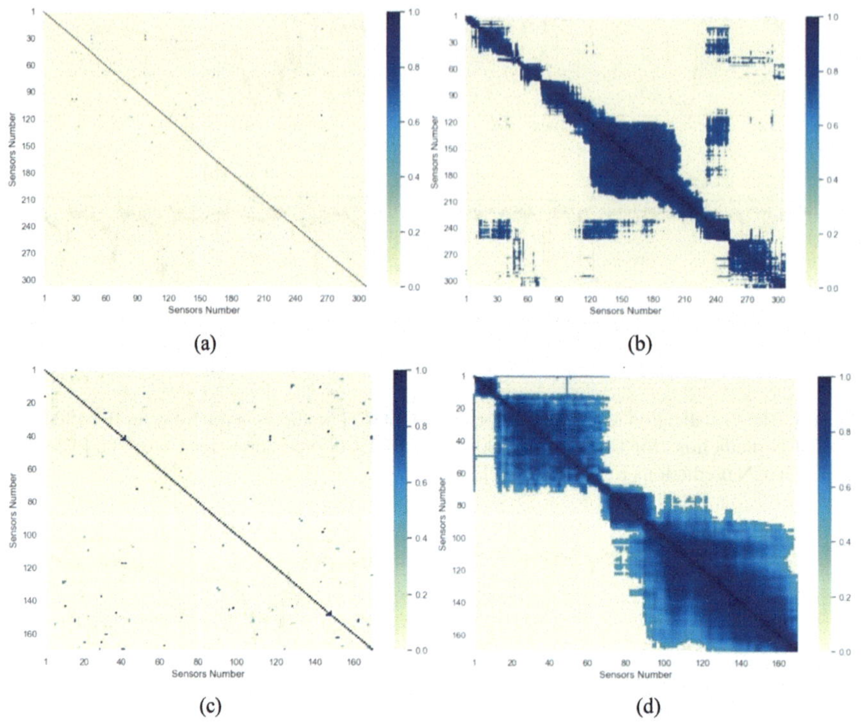

Fig. 4.7 Sensor correlation analysis. (**a**) Correlation based on the adjacency on PEMS04. (**b**) Correlation based on traffic speed on PEMS04. (**c**) Correlation based on the adjacency on PEMS08. (**d**) Correlation based on traffic speed on PEMS08

Fig. 4.7b. This shows that the real topology does not adequately express the spatial correlations of traffic flow. The correlation matrix can further improve the ability of the model to capture global spatial features by reconstructing the correlation between sensors. In this manner, the model can have a better ability to capture global spatial features.

Long-Term Prediction Analysis

The long-time prediction can provide appropriate guidance for those who are preparing to go out. To verify the long-term prediction capability of the model, we adjust the output time step of the model to validation. The results are shown in Table 4.3. As we can see from the table, the model performance decreases slightly when the prediction time step increases. This indicates that T-LGGCN is relatively insensitive to the change of output step length. T-LGGCN can effectively predict traffic in the future to a certain extent.

Table 4.3 Prediction results of different output steps on PEMS04 and PEMS08

	PEMS04				PEMS08			
Steps	RMSE	MAE	Accuracy	R^2	RMSE	MAE	Accuracy	R^2
15 min	2.6374	1.4202	0.9589	0.8975	2.1966	1.1448	0.9657	0.8997
30 min	3.5305	2.0247	0.9450	0.8165	2.9303	1.6418	0.9543	0.8214
45 min	4.1566	2.1695	0.9353	0.7459	3.4309	1.9336	0.9465	0.7548
60 min	4.9374	3.0945	0.9233	0.6417	3.8942	2.2749	0.9392	0.6838

4.3 Urban Road Network Connected Traffic Speed Prediction Model Based on Global Spatiotemporal Characteristics (G-GCGRU)

To analyze the spatial and temporal features of traffic flow separately and construct a prediction model, the model structure consists of three main parts, as illustrated in Fig. 4.8. The first part is the global graph convolutional network module, which performs correlation analysis on the entire feature dataset, computes the correlation matrix, and inputs the correlation matrix and the spatiotemporal matrix with a specified time step into the graph convolutional network. This process extracts the global spatial features of traffic flow data. The second part is the GRU module, where the output from the first part is fed into a GRU network to further analyze the temporal characteristics of the data. The third part is the prediction module, which utilizes the output from the last time step of the GRU network. Through a dense layer, it controls the desired prediction time step and calculates the final prediction results of the model. This establishes the network structure of G-GCGRU.

4.3.1 *Global Spatial Feature Extraction*

The structural characteristics of urban road networks resemble the topological structure of a graph. GCN is a popular neural network for processing graph-structured non-Euclidean data. The topological structure of the road network is represented as an undirected graph $\boldsymbol{G} = (\boldsymbol{V}, \boldsymbol{E})$, where $\boldsymbol{V}$ is the set of nodes representing road segments, and $\boldsymbol{E}$ is the set of edges representing the adjacency between road segments. The computation rules of GCN are as follows:

$$\boldsymbol{H}^{(l+1)} = \sigma\left(\hat{\boldsymbol{A}}\boldsymbol{H}^{(l)}\boldsymbol{W}^{(l)}\right) \tag{4.14}$$

where $\hat{\boldsymbol{A}} = \tilde{\boldsymbol{D}}^{-1/2}\tilde{\boldsymbol{A}}\tilde{\boldsymbol{D}}^{-1/2}$, $\tilde{\boldsymbol{A}} - \boldsymbol{I} + \boldsymbol{A}$, $\boldsymbol{I} \in R^{N \times N}$ is the identity matrix, $\tilde{D}_{ii} = \sum_j \tilde{A}_{ij}$, $\boldsymbol{H}^{(l)}$ denotes the output of the lth layer, $\boldsymbol{W}^{(l)}$ is the weight parameter matrix of the lth layer.

GCN explicitly represents the spatial structure of the actual road network through an adjacency matrix, utilizing $\hat{\boldsymbol{A}}$ for feature convolution. It merges the features of adjacent road segment nodes to obtain a new representation of road segment

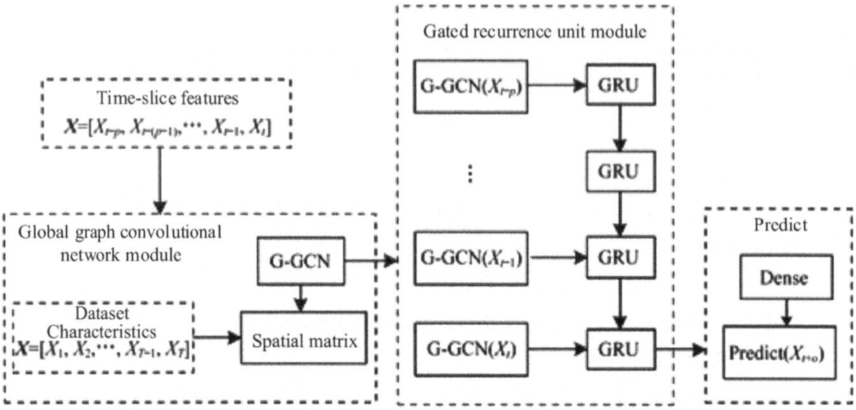

Fig. 4.8 Structure of G-CGCN

features. However, considering the geographical conditions of the roads themselves and the influence of surrounding infrastructure, the spatial impact between some adjacent road segments may be weaker than the spatial impact between non-adjacent road segments. Although the adjacency matrix can express the intuitive adjacency of road segments, it cannot capture the internal spatial influence within the road network. To address this limitation, an implicit representation of the spatial structure within road segments is performed. For any given road segment, the Pearson correlation coefficient method is employed to calculate its correlation with the rest of the road segments in the road network. A certain number, k, of highly correlated road segments are selected to form a correlation matrix $\boldsymbol{C}$. Building on this, leveraging the convolutional concept of GCN, the spatial information of road segments with high correlation is aggregated through $\boldsymbol{C}$ to serve as the new features for the road nodes.

For any two road segments $\boldsymbol{X}_i$ and $\boldsymbol{X}_j$, the method for calculating the correlation matrix is as follows:

$$C_{ij} = \frac{\sum_{t=1}^{T}\left(x_t^i - \overline{X}_i\right)\left(x_t^j - \overline{X}_j\right)}{\sqrt{\sum_{t=1}^{T}\left(x_t^i - \overline{X}_i\right)^2}\sqrt{\sum_{t=1}^{T}\left(x_t^j - \overline{X}_j\right)^2}} \tag{4.15}$$

where $\boldsymbol{X}_i$ represents the feature of traffic flow of road segment i, $\overline{X}_i$ is the mean value of $\boldsymbol{X}_i$.

The relationship between road segment nodes described by the correlation matrix forms a directed weighted graph, where connections between nodes represent influence weights, and the direction signifies the influence relationship. By aggregating information from road segments with high self-correlation through the correlation matrix, the model further enhances the exploration of spatial features. Therefore, the updating rule for the global graph convolutional network is as follows:

$$\boldsymbol{H}^{(l+1)} = \sigma\left(\hat{\boldsymbol{A}}\boldsymbol{H}^{(l)}\boldsymbol{W}^{(l)}\right) \tag{4.16}$$

Inputting the correlation matrix $\boldsymbol{C}$ and the spatiotemporal matrix $\boldsymbol{X}_P^R$ into a two-layer graph convolutional network for spatial feature analysis. The network propagation rules are illustrated in Fig. 4.9. The computation rules are as follows:

$$\mathrm{Ggcn}\left(\boldsymbol{X}_P^R\right) = \mathrm{ReLu}\left(\boldsymbol{C} \cdot \mathrm{ReLu}\left(\boldsymbol{C}\boldsymbol{X}_P^R\boldsymbol{W}^{(0)}\right)\boldsymbol{W}^{(l)}\right) \tag{4.17}$$

4.3.2 *Temporal Feature Extraction*

GRU is one of the mainstream neural networks for handling time series prediction problems, effectively avoiding the issues of vanishing and exploding gradients in RNN. The GRU network consists of three parts: the input layer, hidden layer, and output layer. The key algorithm lies in the computation process of the hidden layer unit block, as illustrated in Fig. 4.10.

For a specific time step, GRU first concatenates the output $\mathbf{h}_{t-1}$ of the previous time step's hidden layer with the current time step's input $\boldsymbol{X}_t$. Through the sigmoid function, the data is transformed into the range [0,1], serving as the reset gate control signal $\boldsymbol{r}_t$ and update gate control signal z_t. Then, utilizing the gate signals, it selectively forgets and retains information from $\mathbf{h}_{t-1}$ and $\boldsymbol{X}_t$. This enables the model to combine the traffic information from the previous time step with the current traffic state, capturing temporal dependencies effectively.

The extracted spatial feature values $\mathrm{Ggcn}\left(\boldsymbol{X}_P^R, \boldsymbol{C}\right)$ are split according to time slices and input into the GRU model for time feature analysis. The computation process is as follows:

$$\boldsymbol{r}_t = \sigma\left(\boldsymbol{W}_r\left[\mathrm{Ggcn}\left(\boldsymbol{X}_P^R, \boldsymbol{C}\right), \boldsymbol{h}_{t-1}\right] + \boldsymbol{b}_r\right) \tag{4.18}$$

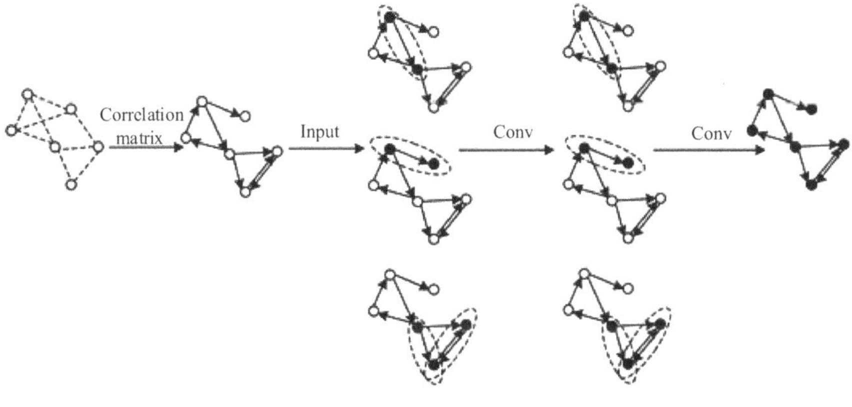

Fig. 4.9 Communication rule of G-GCN

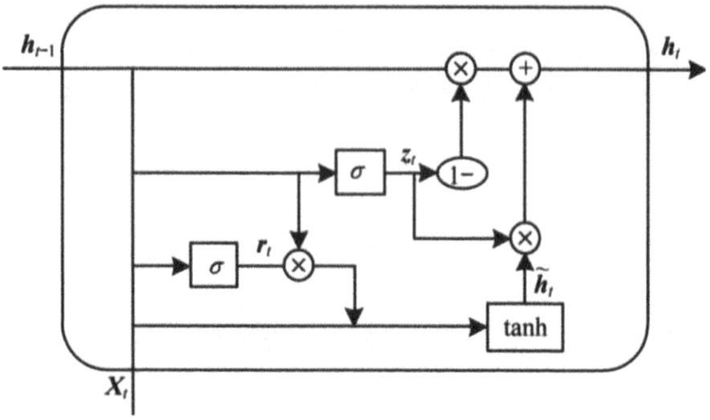

Fig. 4.10 Gated recurrent unit module

$$z_t = \sigma\left(W_z\left[\text{Ggcn}\left(X_P^R, C\right), h_{t-1}\right] + b_z\right) \tag{4.19}$$

$$\tilde{h}_t = \tanh\left(W_{\tilde{h}}\left[\text{Ggcn}\left(X_P^R, C\right), \left(r_t{}^* h_{t-1}\right)\right] + b_{\tilde{h}}\right) \tag{4.20}$$

$$h_t = z_t{}^* h_{t-1} + \left(1 - z_t\right) * \tilde{h}_t \tag{4.21}$$

Obtain the output of the GRU network at the last time step, input it into the dense network, and control the output time step to obtain the final prediction result:

$$y_{t+1} = \text{Dense}\left(h_t\right) \tag{4.22}$$

4.3.3 *Experiments*

4.3.3.1 Experiments Settings

Datasets

This study analyzes road network data from Luohu District, Shenzhen, for the period from January 1, 2015, to January 31, 2015. The dataset comprises information from 156 road segments, aggregated at 15-min intervals, totaling 2976 records of vehicle speed data.

Parameter Settings

The experiment involves setting the number of associated road segments k for a single road segment and constructing the correlation matrix as input to the model. The implementation is based on the TensorFlow framework, with the following main parameter settings: GRU dimensionality is set to 64; learning rate is 0.001; batch size is 32; and the training iterations are set to 3000.

Evaluation Metrics

We use the root mean square error (RMSE), mean absolute error (MAE), coefficient of determination (R^2) as the evaluation metrics, and the detailed definitions are as follows:

$$\mathrm{RMSE}=\sqrt{\frac{1}{PR}\sum_{p=1}^{P}\sum_{r=1}^{R}\left(y_p^r-y_p^{r'}\right)^2} \tag{4.23}$$

$$\mathrm{MAE}=\frac{1}{PR}\sum_{p=1}^{P}\sum_{r=1}^{R}\left|y_p^r-y_p^{r'}\right| \tag{4.24}$$

$$R^2=1-\frac{\sum_{p=1}^{P}\sum_{r=1}^{R}\left(y_p^r-y_p^{r'}\right)^2}{\sum_{p=1}^{P}\sum_{r=1}^{R}\left(y_p^r-\bar{y}\right)^2} \tag{4.25}$$

where P represents the input time steps, R is the numbers of road segment, y_p^r and $y_p^{r'}$ denote the real traffic speed data and predicted data, respectively.

4.3.3.2 Experiments Results

Analysis of the Number of Associated Road Segments k

The G-GCGRU model considers the spatial correlation of road segments in the entire road network. By analyzing the correlation of vehicle speeds between road segments in the road network, correlation values for each pair of road segments are obtained, as shown in Fig. 4.11. Figure 4.11a illustrates the correlation between adjacent road segments in the road network, while Fig. 4.11b presents the situation of highly correlated road segments calculated based on the associated segment quantity k. Darker colors in Fig. 4.11b indicate higher correlation values. It can be observed that the node correlation in Fig. 4.11a is significantly weaker than in Fig. 4.11b, indicating that the expression of spatial correlations in the road network is not fully captured by the actual topological structure. The correlation matrix, by reconstructing the correlations between road segments, deeply analyzes highly correlated segments and discards information from less correlated segments, thereby further enhancing the model's ability to capture spatial features.

The associated segment quantity k is an important factor influencing the predictive performance of the model. The relationship between k and RMSE, MAE values is depicted in Fig. 4.12. Overall, as the value of k increases, the model error shows a decreasing trend, indicating that aggregating spatial information from multiple road segments is beneficial for improving the model's ability to explore spatial features. When k is set to 6, the model error reaches its minimum. Subsequently, as k increases, the model error slowly grows, suggesting that 6 is the optimal number of

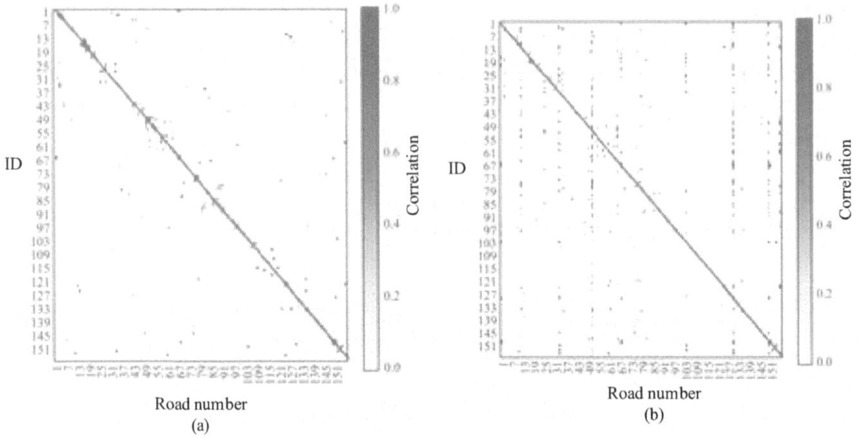

Fig. 4.11 Road correlation matrix comparation. (**a**) Adjacent roadway correlation matrix. (**b**) Correlation matrix for highly correlated road sections

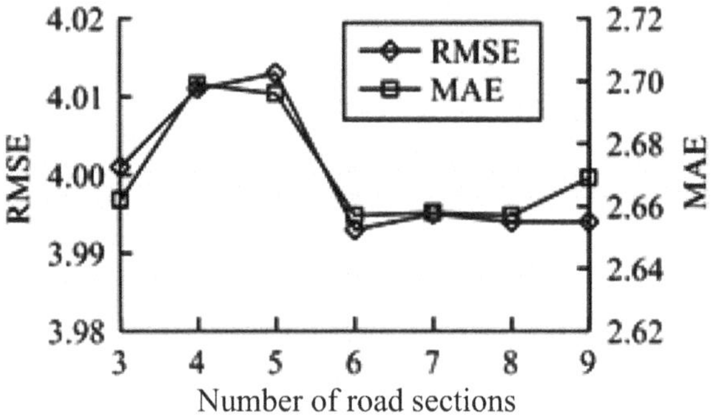

Fig. 4.12 Analysis of correlation road number

associated road segments. Moreover, an excessive k introduces additional data information, which can negatively impact the model's predictive capability.

Comparative Experiment

To verify the effectiveness and superiority of the proposed model, a comparison was conducted with GCN, GRU, and the hybrid GCN-GRU model. The results are presented in Table 4.4. It can be observed that among these four prediction models, the performance of the proposed G-GCGRU model is superior.

Figure 4.13 shows the fitting curve for the daily data of road segment with the ID 102129 in the road network for different models. It can be observed that the proposed model has a relatively good fitting effect. Compared to the GRU network, the prediction performance of the GCN network is significantly inferior, indicating that

Table 4.4 Model performance comparison

Model	RMSE	MAE	R^2
GCN	5.347	3.929	0.627
GRU	4.193	2.783	0.841
GCN-GRU	4.079	2.736	0.849
G-GCGRU	3.996	2.663	0.853

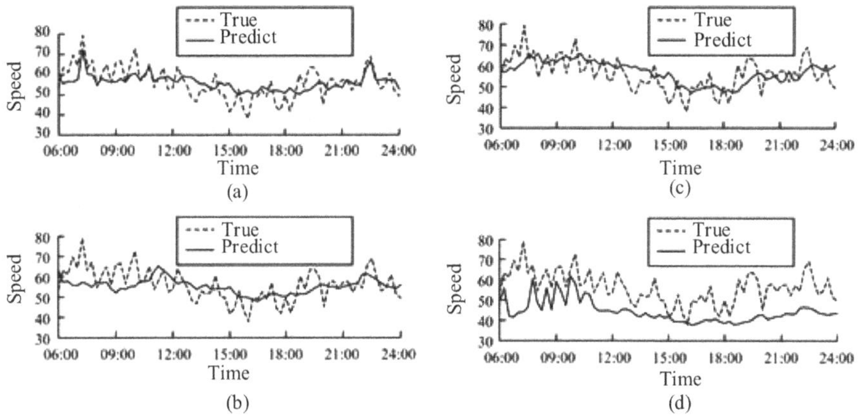

Fig. 4.13 Prediction results comparison of four models. (**a**) G-GCGRU. (**b**) GCN-GRU. (**c**) GRU. (**d**) GCN

time features are crucial for traffic flow. In comparison to both GRU and GCN, the GCN-GRU model shows a certain improvement in prediction accuracy, suggesting that the hybrid network model comprehensively explores both temporal and spatial features, conducting a more comprehensive feature extraction. However, when compared to G-GCGRU, the performance of the GCN-GRU model is slightly lower, indicating that the convolutional effect of the correlation matrix on spatial features is superior to the adjacency matrix, effectively enhancing the model's prediction accuracy.

Multi-Step Prediction Capability

To validate the multi-step prediction capability of the model proposed in this paper, adjustments were made to the model's output time step and various parameters for verification. The results are shown in Fig. 4.14. It can be observed that with the increase in the prediction time step, the model's performance slightly decreases. However, the predictions for each time step show minimal fluctuations, indicating that the performance of the proposed model is relatively insensitive to changes in the output step length and can effectively predict the traffic state for longer future time periods.

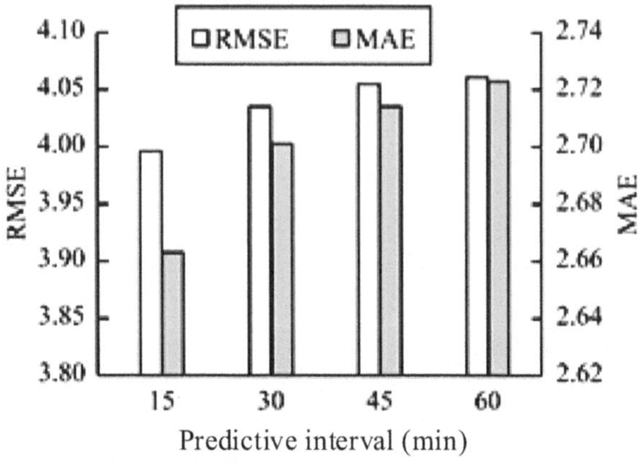

Fig. 4.14 Multi-step prediction result

4.4 PLU-MCN: Perturbation Learning Enhanced U-Shaped Multi-Graph Convolutional Network

The proposed PLU-MCN is shown in Fig. 4.15, which consists of a Perturbation Learning (PL) module and a U-shaped multi-graph convolutional network (U-MCN). Among them, the PL module includes the perturbation learning encoder (PLE) and the perturbation learning decoder (PLD) module; U-MCN consists of the multi-graph convolutional encoder (MGCE) module, the multi-graph convolutional decoder (MGCD) module, and the neck module combined according to the U-shaped structure.

4.4.1 Perturbation Learning (PL) Module

The perturbation data, such as missing values and outliers in the traffic flow dataset, has an essential impact on the model's prediction performance. Thus, data preprocessing is necessary. This paper proposes a novel PL module to avoid repetitive data preprocessing in the same dataset type. The role of the PL module is to learn the perturbation-resistant features from the perturbation data. Perturbation-resistant features contain key traffic flow features in the perturbation dataset and can be used as inputs to the model. The PL module is designed based on the encoder and decoder architecture, consisting of the PLE and PLD modules. When training the PLU-MCN, this paper needs to pre-train the PL module first, and the inputs and outputs of the PL module are datasets with and without perturbations, respectively. After the pre-training is completed, the weights of the PLE module are fixed, and the perturbation-resistant features output from the PLE module is used as the input of the U-MCN module.

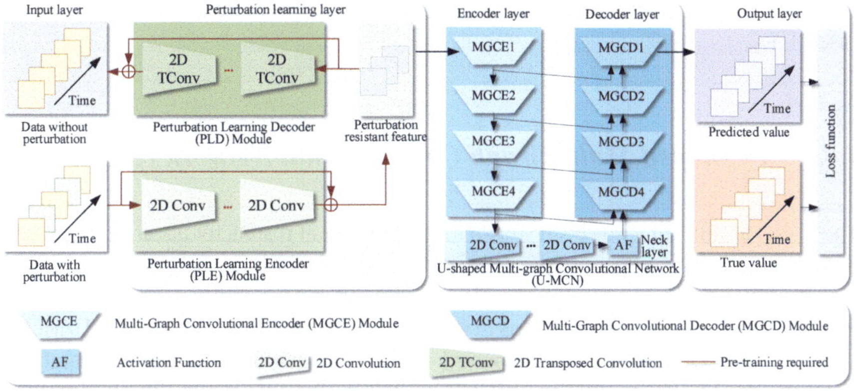

Fig. 4.15 Structure of PLU-MCN

The PLE module consists of 2D convolution and residual connection, and the 2D convolution operation $f_{2\mathrm{DConv}}$ is defined as in Eq. (4.26).

$$\boldsymbol{X}_{2\mathrm{DConv}}^{l+1} = f_{\mathrm{AF}}\left(\boldsymbol{W}_{2\mathrm{DConv}}^{l} {}^{*} \boldsymbol{X}_{2\mathrm{DConv}}^{l}\right) \tag{4.26}$$

where $\boldsymbol{X}_{2\mathrm{DConv}}^{l+1}$ and $\boldsymbol{W}_{2\mathrm{DConv}}^{l}$ are the output and input of the l-th layer 2D convolution operation, respectively, "*" is the convolution operation, f_{AF} is the activation function, and $\boldsymbol{W}_{2\mathrm{DConv}}^{l}$ is the weight of the l-th layer 2D convolution operation.

Given the input of the PLE module is $\boldsymbol{X}_{\mathrm{En}} \in R^{N \times S}$, the output $\boldsymbol{X}_{\mathrm{En1}} \in R^{N \times K}$ is obtained after multiple 2D convolutions, the residual join is introduced to avoid feature loss, and the operation is defined as shown in Eq. (4.27):

$$\boldsymbol{X}_{\mathrm{h}} = f_{\mathrm{AF}}\left(\boldsymbol{X}_{\mathrm{En1}} + f_{2\mathrm{DConv}}\left(\boldsymbol{X}_{\mathrm{En}}\right)\right) \tag{4.27}$$

where $X_{\mathrm{h}} \in R^{N \times K}$ is the final output value of the PLE module, which is the perturbation-resistant feature and the input to the PLD module.

The PLD module consists of 2D transposed convolution and residual connection for decoding the perturbation-resistant feature. The 2D transposed convolution $f_{2\mathrm{DTConv}}$ is defined as Eq. (4.28). Unlike $f_{2\mathrm{DConv}}$ which reduces the dimension of features, $f_{2\mathrm{DTConv}}$ is used to increase the dimension of features.

$$\boldsymbol{X}_{2\mathrm{DTConv}}^{l+1} = f_{\mathrm{AF}}\left(\boldsymbol{W}_{2\mathrm{DTConv}}^{l} {}^{*} \boldsymbol{X}_{2\mathrm{DTConv}}^{l}\right) \tag{4.28}$$

where $\boldsymbol{X}_{2\mathrm{DTConv}}^{l+1}$ and $\boldsymbol{W}_{2\mathrm{DTConv}}^{l}$ are the output and input of the l-th layer 2D transposed convolution operation, respectively, and $\boldsymbol{W}_{2\mathrm{DTConv}}^{l}$ is the weight of the l-th layer 2D transposed convolution operation. After multi-layer 2D transposed convolution, the output $\boldsymbol{X}_{\mathrm{De1}} \in R^{N \times S}$ is obtained, and the residual connection is introduced to avoid feature loss, and the operation is defined as Eq. (4.29).

$$\boldsymbol{X}_{\mathrm{De}} = f_{\mathrm{AF}}\left(\boldsymbol{X}_{\mathrm{h}} + f_{2\mathrm{DTConv}}\left(\boldsymbol{X}_{\mathrm{De1}}\right)\right) \tag{4.29}$$

where $X_{De} \in R^{N \times S}$ is the final output value of the PLD module and the unperturbed case of X_{En}.

4.4.2 U-Shaped Multi-Graph Convolutional Network (U-MCN)

The U-Net architecture has been shown to have excellent results for accurate extraction of features [1]. This paper adopts the U-shaped architecture to construct the U-MCN model, and U-MCN consists of MGCE modules, MGCD modules, and neck module. Five MGCE modules are connected in series to build the encoder part of the U-MCN, and five MGCD modules are connected in series to build the decoder part of the U-MCN. The MGCE module and MGCD module perform the down-sampling and up-sampling operations on the features, respectively, and the structure is shown in Fig. 4.16.

4.4.2.1 Multi-Graph Convolutional Encoder (MGCE) Module

The MGCE module consists of the dilated convolution module, BatchNorm, multi-graph convolution module, 2D convolution, and residual connection. The dilated convolution module is used to expand the range of the sensory field of the convolution kernel, and the dilated causal convolution operation f_{DCN} is defined as shown in Eq. (4.30).

$$X_{DCN}^{l+1} = f_{AF}\left(W_{DCN}^{l} * X_{DCN}^{l}\right) \tag{4.30}$$

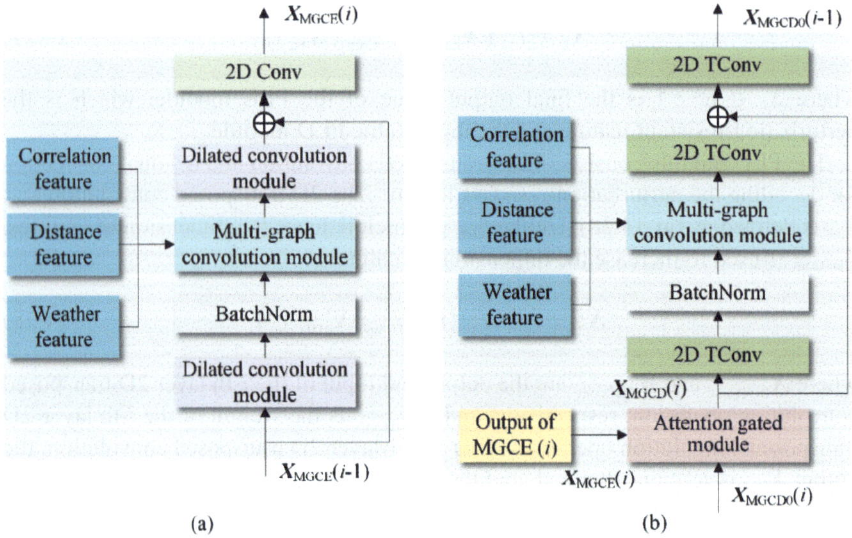

Fig. 4.16 Structure of the MGCE module and the MGCD module. (**a**) Multi-graph convolutional encoder (MGCE) module. (**b**) Multi-graph convolutional decoder (MGCD) module

where $\boldsymbol{X}_{\mathrm{DCN}}^{l+1}$ and $\boldsymbol{X}_{\mathrm{DCN}}^{l}$ are the ouput and input of the l-th layer of the dilated convolution module, respectively, $\boldsymbol{W}_{\mathrm{DCN}}^{l}$ is the weight of the l-th layer of the dilated convolution module, and the dilation rate of the dilation convolution is k_{d}. The BatchNorm is used to keep the inputs of each layer of the neural network identically distributed.

Multi-graph convolution module is constructed for obtaining spatial features of traffic flow under the influence of data correlation, distance, and weather. From Sect. 4.3.2, the node's correlation feature matrix $\boldsymbol{A}_{\mathrm{C}}$, distance feature matrix $\boldsymbol{A}_{\mathrm{D}}$, and weather feature matrix $\boldsymbol{A}_{\mathrm{W}}$ are obtained. The multi-graph convolution operation f_{MGCN} is defined as shown in Eq. (4.31).

$$\boldsymbol{X}_{\mathrm{GCN}}^{l+1} = f_{\mathrm{AF}}\left(\mathrm{FC}_1\left(\boldsymbol{A}_{\mathrm{C}}\boldsymbol{X}_{\mathrm{GCN}}^{l}\boldsymbol{W}_{\mathrm{GCN}}\right) + \mathrm{FC}_2\left(\boldsymbol{A}_{\mathrm{D}}\boldsymbol{X}_{\mathrm{GCN}}^{l}\boldsymbol{W}_{\mathrm{GCN}}\right) + \mathrm{FC}_3\left(\boldsymbol{A}_{\mathrm{W}}\boldsymbol{X}_{\mathrm{GCN}}^{l}\boldsymbol{W}_{\mathrm{GCN}}\right)\right) \quad (4.31)$$

where $\boldsymbol{X}_{\mathrm{GCN}}^{l+1}$ and $\boldsymbol{X}_{\mathrm{GCN}}^{l+1}$ are the output and input of the multi-graph convolution module of the l-th layer, respectively, and FC(•) is a fully connected operation for obtaining the weights of the multi-graph features.

In the MGCE module, the role of the dilated convolution module is to obtain long-term temporal feature and the role of the multi-graph convolution module is to obtain spatial feature. The operation of the MGCE module is shown in Eq. (4.32).

$$\boldsymbol{X}_{\mathrm{MGCE}}(i+1) = f_{\mathrm{2DConv}}\left(f_{\mathrm{2DConv}}\left(\boldsymbol{X}_{\mathrm{MGCE}}(i)\right) + f_{\mathrm{DCN}}\left(f_{\mathrm{MGCN}}\left(f_{\mathrm{DCN}}\left(\boldsymbol{X}_{\mathrm{MGCE}}(i)\right)\right)\right)\right) \quad (4.32)$$

where $\boldsymbol{X}_{\mathrm{MGCE}}(i)$ and $\boldsymbol{X}_{\mathrm{MGCE}}(i + 1)$ are the $(i + 1)$-th MGCE module's input and output, respectively. The MGCE module is used to perform down-sampling operations on the features.

4.4.2.2 Multi-Graph Convolutional Decoder (MGCD) Module

The MGCD module is used for up-sampling the features. Unlike the MGCE module, the input of the i-th MGCD module is obtained by fusing the outputs of the $(i + 1)$-th MGCD module and the i-th MGCE module. The attention gated module is designed to fuse the outputs of the i-th MGCE module and the $(i + 1)$-th MGCD module, as shown in Fig. 4.17. The attention gated module operation is defined as Eq. (4.33).

$$\begin{gathered}\boldsymbol{X}_{\mathrm{MGCD}}(i) = f_{\mathrm{Att}}\left(\boldsymbol{X}_{\mathrm{MGCD0}}(i)\right) \otimes \sigma\left(f_{\mathrm{Att}}\left(\boldsymbol{X}_{\mathrm{MGCE}}(i)\right)\right) \\ f_{\mathrm{Att}}(\boldsymbol{X}) = (\boldsymbol{X})^{\mathrm{Tr}} \cdot \frac{\mathrm{e}^{\left(X \cdot (X)^{\mathrm{Tr}}\right)}}{\sum_1^T \mathrm{e}^{\left(X \cdot (X)^{\mathrm{Tr}}\right)}}\end{gathered} \quad (4.33)$$

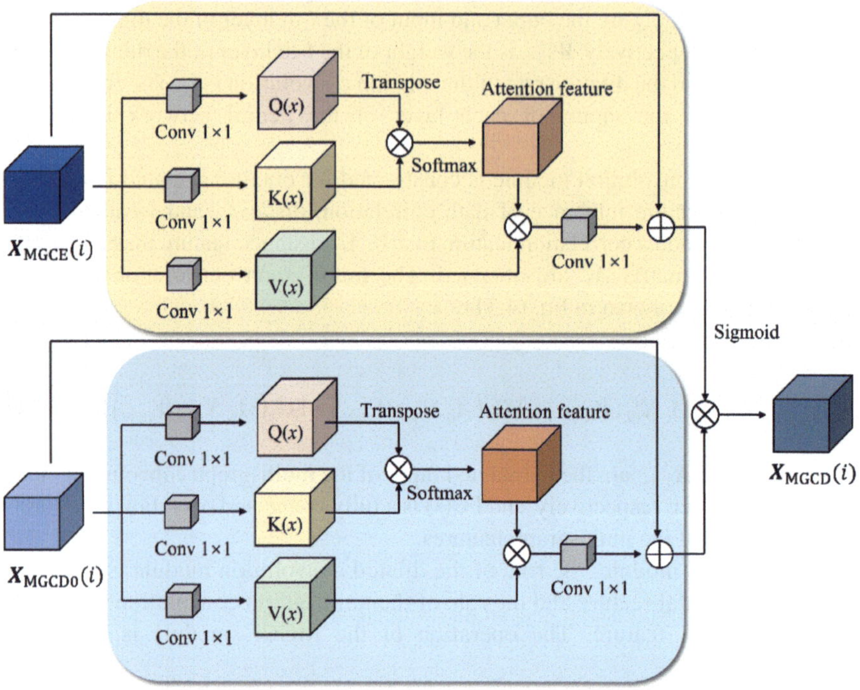

Fig. 4.17 Attention gated module

where $\boldsymbol{X}_{\mathrm{MGCD0}}(i)$ and $\boldsymbol{X}_{\mathrm{MGCD}}(i)$ are the output of the $(i + 1)$-th MGCD module and the fusion input of the i-th MGCD module, respectively. $\boldsymbol{X}_{\mathrm{MGCE}}(i)$ is the output of the i-th MGCE module, $\otimes$ is the Hadamard product, Tr is the transpose operation, σ is the Sigmoid activation function. The MGCD module is defined in Eq. (4.34). The first MGCD module finally outputs the predicted value of traffic flow.

$$\boldsymbol{X}_{\mathrm{MGCD0}}(i-1) = f_{\mathrm{2DTConv}}\left(f_{\mathrm{2DTConv}}\left(\boldsymbol{X}_{\mathrm{MGCD}}(i)\right) + f_{\mathrm{2DTConv}}\left(f_{\mathrm{MGCN}}\left(f_{\mathrm{2DTConv}}\left(\boldsymbol{X}_{\mathrm{MGCD}}(i)\right)\right)\right)\right) \tag{4.34}$$

4.4.2.3 Neck Module

The neck module is designed to connect the MGCE and MGCD modules and consists mainly of 2D convolutional operations. The input and output of the neck module are the output of the fourth MGCE module and the input of the fourth MGCD module, respectively.

4.4.3 PLU-MCN Training Process

The training process of PLU-MCN is divided into two stages: pre-training and formal training. In the pre-training stage, only the PL module is involved in weight updating among all the modules of PLU-MCN. The input of PLU-MCN is the data with perturbation, and the output is the data without perturbation in the corresponding period. After the pre-training, the formal training phase is entered. In the formal training stage, the output of the PLE module is used as the input of the U-MCN model, the weights of the PLE module are kept unchanged, and the PLD module is not involved in training. The loss function in both phases is the mean square error (MSE), which is defined as follows:

$$\mathrm{MSE} = \frac{\sum_{i=1}^{p} \left| \boldsymbol{Y}_{\mathrm{R}}(i) - \boldsymbol{Y}_{\mathrm{P}}(i) \right|^{2}}{p} \tag{4.35}$$

where $\boldsymbol{Y}_{\mathrm{R}}$ and $\boldsymbol{Y}_{\mathrm{P}}$ is the true value and predicted value of traffic flow, and p is the number of traffic flow training samples. The specific training process of the PLU-MCN model is shown in Algorithm 1.

Algorithm 1 PLU-MCN Training Algorithm

Input: Model input with perturbation: $\boldsymbol{H}_{\mathrm{NI}} \in R^{N \times S}$, model input without perturbation: $\boldsymbol{H}_{\mathrm{I}} \in R^{N \times S}$, correlation feature matrix $\boldsymbol{A}_{\mathrm{C}} \in R^{N \times N}$, distance feature matrix $\boldsymbol{A}_{\mathrm{D}} \in R^{N \times N}$, weather feature matrix $\boldsymbol{A}_{\mathrm{W}} \in R^{N \times N}$, dimension of perturbation-resistant feature $N \times K$

Output: Learned PLU-MCN model

// Initialize parameters
PL {PLE ($\boldsymbol{\theta}_1$), PLD ($\boldsymbol{\theta}_2$)} //Initialize PL module
//Pre-training stage
for epoch in max epoch **do**
    $\boldsymbol{H}_{\mathrm{NO}} \in R^{N \times S} = \mathrm{PL}(\boldsymbol{H}_{\mathrm{NI}})$
    Minimize loss value MSE ($\boldsymbol{H}_{\mathrm{NO}}$, $\boldsymbol{H}_{\mathrm{I}}$)
    Update PL {PLE ($\boldsymbol{\theta}_1$), PLD ($\boldsymbol{\theta}_2$)}
end for
//Formal training stage
PLE ($\boldsymbol{\theta}_1$) // Fix the weights of PLE module
U-MCN ($\boldsymbol{\theta}_3$) //Initialize U-MCN
for epoch in max epoch **do**
    $\boldsymbol{X}_{\mathrm{h}} \in R^{N \times K} = \mathrm{PLE}(\boldsymbol{H}_{\mathrm{NI}})$ // Perturbation-resistant feature
    $\boldsymbol{Y}_{\mathrm{P}} = \text{U-MCN}(\boldsymbol{X}_{\mathrm{h}})$ // Predictive value
    Minimize loss value MSE ($\boldsymbol{Y}_{\mathrm{P}}$, $\boldsymbol{Y}_{\mathrm{T}}$)
    Update U-MCN ($\boldsymbol{\theta}_3$)
end for
Output the learned PLU-MCN model

4.4.4 Experiments

4.4.4.1 Datasets Description and Preprocessing

This paper uses five publicly available real-world datasets as test data for models, namely PEMS03, PEMS04, PEMS07, PEMS08, and SZ-TAXI [2, 3]. The PEMS (03–08) dataset is collected from Caltrans data on California highways in the USA, the collection interval is 5 min, the data types include traffic flow, speed, and occupancy, and the traffic flow is used as the test data of PEMS (03–08). The SZ-TAXI dataset is collected from taxi data in Shenzhen, China at the interval of 15 min, and the data type is traffic speed. Table 4.5 shows the details of the five public datasets. In this paper, in addition to collecting the data of traffic flow and traffic speed type, we also need to collect the weather status of the area where the dataset is located, and the data source is "http://tianqi.2345.com/."

Due to the fact that both the PEMS (03–08) and SZ-TAXI datasets have been preprocessed, this paper generates five datasets corresponding to the presence of perturbations by adding the random noise of the Gaussian distribution $N_{\mathrm{Gau}} \in (0, \mathrm{gau}^2)$ to the datasets: the PEMS03-G, the PEMS04-G, the PEMS07-G, the PEMS08-G, and the SZ-TAXI-G.

4.4.4.2 Experimental Settings

Hyperparameter Settings

This paper develops PLU-MCN model based on Pytorch-GPU, and conFig.s the experimental environment based on Windows 11 operation system, Intel i7 processor, and Nvidia 3080 graphics card. PLU-MCN is optimized by Adam, with the initial learning rate is 0.01, the learning decay rate is 0.2, the batch size is 30, and the maximal number of iterations is 800. This paper divides the training set, validation set and test set in the ratio of 7:1:2, and repeats 15 rounds to output the average test results. S, K in PL module is set to 12, 10 respectively. The predicted sequence length M of PLU-MCN is 12.

Table 4.5 Information about the datasets

Datasets	PEMS03	PEMS04	PEMS07	PEMS08	SZ-TAXI
Location	In California, USA	In California, USA	In California, USA	In California, USA	In Shenzhen, China
Date	9/1–11/30/2018	1/1–2/28/2018	5/1–8/31/2017	7/1–8/31/2016	1/1–1/31/2015
Time interval	5 min	5 min	5 min	5 min	15 min
Data type	Flow	Flow	Flow	Flow	Speed
Nodes	358	307	883	170	156
Intervals	26,208	16,992	28,224	17,856	2976

Baseline and State of the Art Models

This paper compares PLU-MCN with the following 14 baselines belonging to four classes. (1) Time series prediction models: ARIMA [4], LSTM [5]. (2) Spatial feature extraction model: CNN [6]. (3) GCN-based models: DCRNN [7], STGCN [8], ASTGCN [9], STSGCN [3], STFGNN [10], and Auto-DSTSGN [11]. (4) Attention-based models: STTN [12], GMAN [13], TFormer [14], ASTGNN [15], PDForme [16]. ARIMA is a classical model-driven time series prediction method. LSTM and CNN are two typical deep learning methods based single-feature extraction. The other 11 models are deep learning methods that make unique improvements in spatial-temporal feature extraction.

Evaluation Indicators

This paper uses three common metrics in traffic flow prediction tasks, i.e., root mean square error (RMSE), mean absolute error (MAE), and mean absolute percentage error (MAPE), to evaluate the prediction accuracy of the proposed model [13]. This paper also uses Params and FLOPs to measure the spatial and temporal complexity of models, respectively.

$$\mathrm{RMSE} = \sqrt{\frac{1}{p}\sum_{i=1}^{p}\left(\boldsymbol{Y}_{\mathrm{R}}(i) - \boldsymbol{Y}_{\mathrm{P}}(i)\right)^2} \tag{4.36}$$

$$\mathrm{MAE} = \frac{1}{p}\sum_{i=1}^{p}\left|\boldsymbol{Y}_{\mathrm{R}}(i) - \boldsymbol{Y}_{\mathrm{P}}(i)\right| \tag{4.37}$$

$$\mathrm{MAPE} = \frac{100\%}{p}\sum_{i=1}^{p}\left|\frac{\boldsymbol{Y}_{\mathrm{R}}(i) - \boldsymbol{Y}_{\mathrm{P}}(i)}{\boldsymbol{Y}_{\mathrm{R}}(i)}\right| \tag{4.38}$$

4.4.4.3 Experimental Results

Predictive Performance of PLU-MCN

This paper uses the test set data of PEMS (03–08) and SZ-TAX to demonstrate the prediction performance of PLU-MCN. As shown in Fig. 4.18, PLU-MCN can fit the traffic flow of the selected dataset with better prediction performance. The analysis shows that the PEMS (03–08) dataset reaches the peak traffic flow around 6:30 am. To verify the convergence effect of PLU-MCN in the five datasets, this paper shows the iteration effect of PLU-MCN in the training set. The RMSE iteration results of the training set after normalization are shown in Fig. 4.19. PLU-MCN reaches convergence at about 90 rounds of iterations, then the convergence is the fastest before the 18th round of iterations, and the convergence is gradually slowed down from the 18th to the 90th round of iterations.

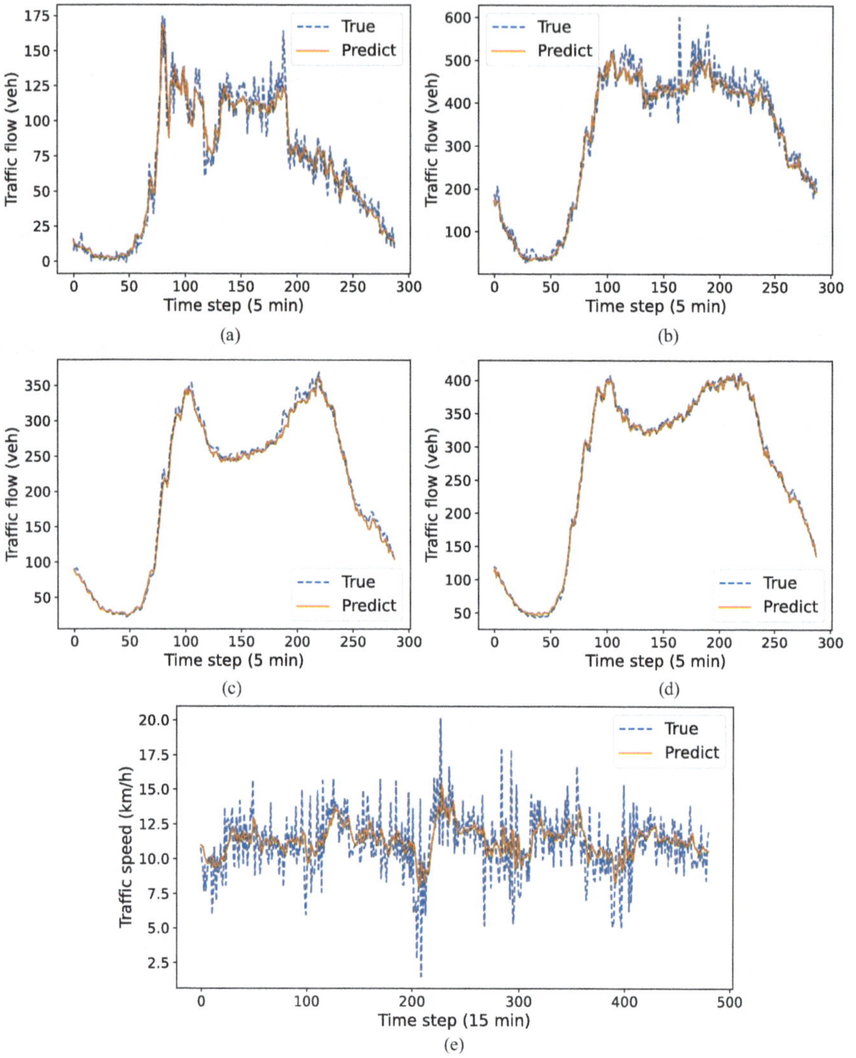

Fig. 4.18 Fitting performance of PLU-MCN. (**a**) PEMS03 (RMSE = 9.290). (**b**) PEMS04 (RMSE = 29.419). (**c**) PEMS07 (RMSE = 7.658). (**d**) PEMS08 (RMSE = 5.543). (**e**) SZ-TAXI (RMSE = 2.079)

Prediction Results of PLU-MCN and Comparison Models

This paper uses PEMS (03–08) to demonstrate the average predictive performance of the comparison models in the first three categories, and the results are shown in Table 4.6. With the evaluation metric of RMSE, the prediction performance of PLU-MCN is improved by 8.237% compared to the state-of-the-art model. To verify the effect of perturbation data on model prediction, this paper demonstrates the variation of RMSE accuracy about different models under PEMS03-G, PEMS04-G,

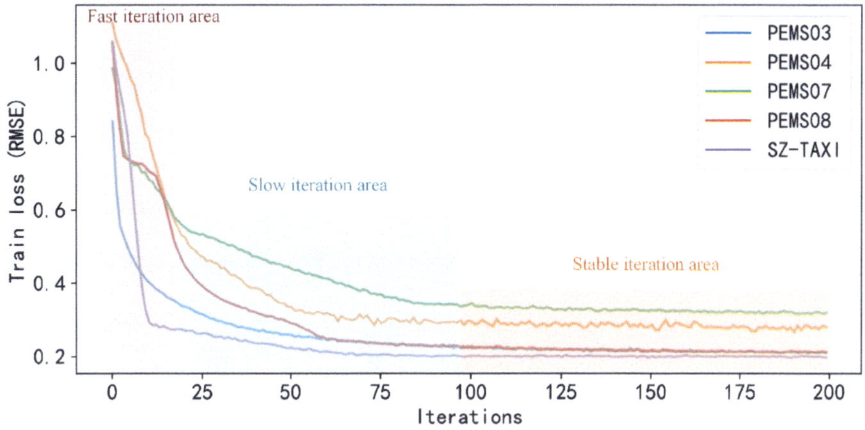

Fig. 4.19 Convergence performance of PLU-MCN

PEMS07-G, and PEMS08-G datasets, as shown in Fig. 4.20. It can be found that with the increase of the Gaussian noise standard deviation gau, the PLU-MCN proposed in this paper is minimized by the noise and has good prediction performance.

Accurately capturing the traffic flow during peak hours is the critical basis for verifying the effectiveness of models. This paper selects the SZ-TAXI and PEMS08 datasets as test samples to validate the model's prediction performance during peak hours, as shown in Fig. 4.21. This paper selects two key hours of data, 8:00 a.m. to 10:00 a.m. and 5:00 p.m. to 7:00 p.m., as the test. The experimental results show that PLU-MCN has better prediction performance and can better sense the traffic flow during peak hours.

Accurate prediction of holiday traffic is one of the key metrics for model performance. This paper chooses PEMS04 as a test set to show the prediction results of different models on February 19 (President's Day), as shown in Fig. 4.22. It can be found that the PLU-MCN proposed in this paper has better prediction performance compared to several classical attention-based prediction models.

The long-term prediction performance of models is crucial in practical applications. This paper divides the prediction period into 5, 15, 30, and 60 min to validate the long-term prediction performance of the model. PEMS03 is used as a test set to demonstrate the prediction results of different models, as shown in Fig. 4.23. As the prediction period increases, the prediction performance of the model gradually decreases, and the PLU-MCN proposed in this paper always has the optimal prediction accuracy.

Ablation Experiments

This paper proposes three variants to validate the structural superiority of PLU-MCN. PLU-MCN-NPL represents the removal of the PL module. PLU-MCN-A represents the use of adjacency matrix instead of correlation feature matrix, distance

Table 4.6 Demonstration of average prediction accuracy

Classical models											
Datasets	Metrics	ARIMA	LSTM	CNN	DCRNN	STGCN	ASTGCN	STSGCN	STFGNN	Auto-DSTSGN	PLU-MCN
PeMS03	RMSE	41.587	35.233	32.987	30.427	30.314	29.676	28.873	28.342	25.170	**20.974**
	MAE	26.896	21.132	20.725	18.239	17.485	17.693	17.379	16.773	**14.592**	15.153
	MAPE (%)	27.314	21.561	20.773	18.924	17.472	19.289	16.882	16.309	14.226	**13.005**
PeMS04	RMSE	57.139	41.732	39.076	38.383	35.423	35.222	34.136	31.891	30.488	**28.877**
	MAE	37.311	27.162	27.078	24.926	22.721	22.933	21.014	19.836	18.854	**18.211**
	MAPE (%)	26.050	18.598	17.623	17.482	14.688	16.577	13.905	13.021	13.213	**12.178**
PeMS07	RMSE	55.348	45.927	40.794	38.856	38.758	42.571	38.894	35.878	33.024	**31.705**
	MAE	37.849	30.021	32.141	25.764	25.382	28.352	24.768	22.087	20.089	**18.449**
	MAPE (%)	17.366	13.532	12.361	11.693	11.087	13.889	10.663	9.253	8.571	**7.690**
PeMS08	RMSE	41.671	34.074	29.528	27.837	27.829	28.380	27.555	26.231	23.772	**22.108**
	MAE	28.123	22.373	21.858	17.861	18.035	18.591	17.883	16.642	14.749	**13.989**
	MAPE (%)	18.259	14.682	13.790	11.458	11.427	13.246	11.709	10.608	9.451	**8.733**

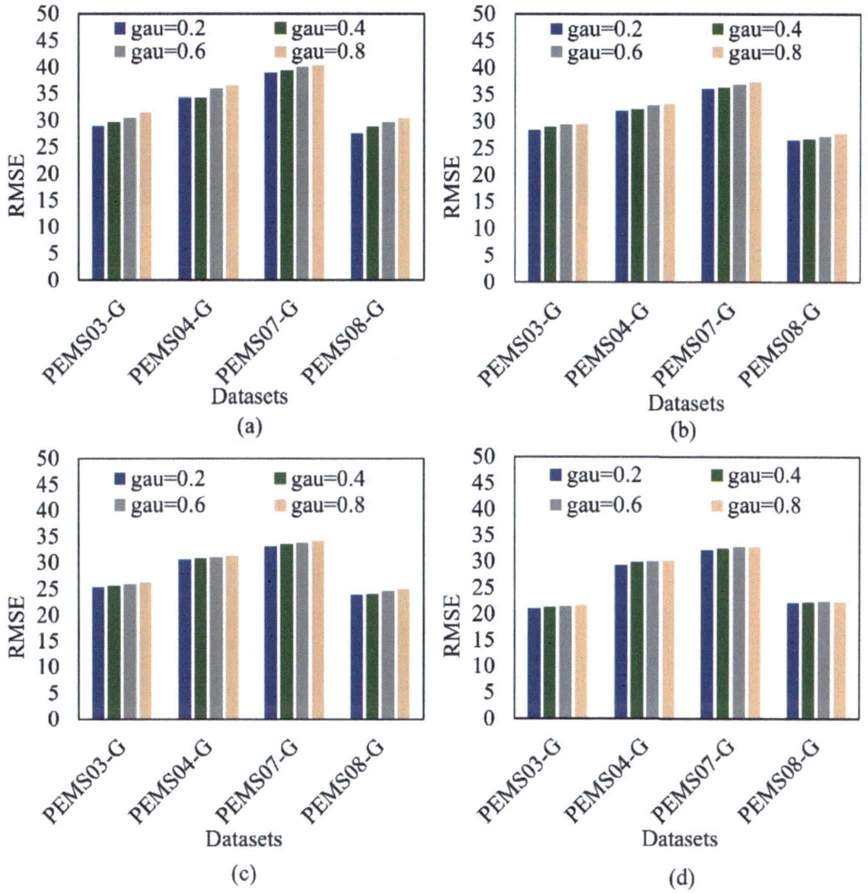

Fig. 4.20 Effect of noise on prediction performance. (**a**) STSGCN. (**b**) STFGCN. (**c**) Auto-DSTSGN. (**d**) PLU-MCN

feature matrix, and weather feature matrix in the multi-graph convolution module. PLU-MCN-NA represents the removal of the attention mechanism in the MGCD module. The prediction performance of the three variants of the model using PEMS03-G (gau = 0.2) and PEMS07-G (gau = 0.2) as test data is shown in Fig. 4.24. Taking PEMS03 (gau = 0.2) as an example, the RMSE of PLU-MCN is reduced by 16.7% compared to PLU-MCN-NA, but the FLOPS and Params are improved by 87.648% and 49.773%, respectively.

This paper also tests the scalability of the PL module to other models using PEMS (03–08)-G (gau = 0.8) as the test set. The PL module is applied to STSGCN, STFGNN, and Auto-DSTSGN, respectively. The experimental results are shown in Fig. 4.25. It can be seen that the addition of the PL module improves the prediction performance of all four models in the dataset with perturbations.

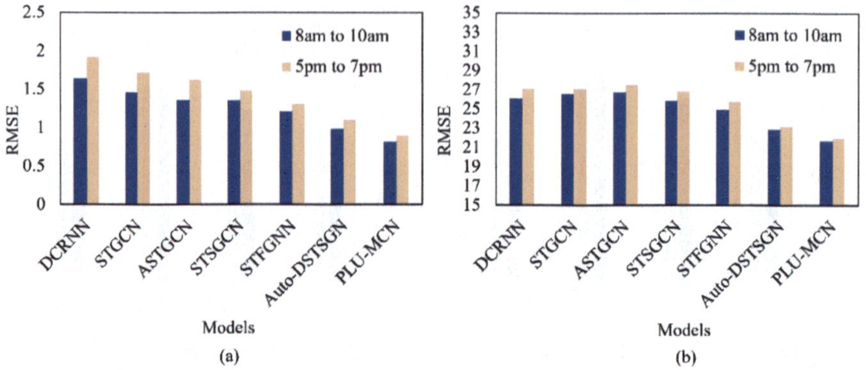

Fig. 4.21 Predictive performance during peak hours. (**a**) SZ-TAXI. (**b**) PEMS08

Fig. 4.22 Predictive performance during holidays

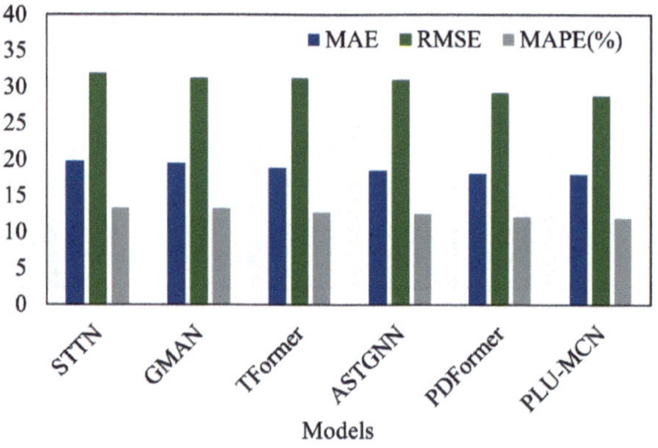

4.4.4.4 Discussion

Accurately fitting the traffic flow of different regions best reflects the practicality of the traffic flow prediction model. This paper uses test set data from PEMS (03–08) and SZ-TAX to demonstrate the predictive performance of PLU-MCN. Figure 4.18 shows that PLU-MCN can well fit the traffic flow of five real-world datasets. It can be observed that the PEMS (03–08) dataset reaches its peak traffic around 6:30 am. By analyzing the RMSE iteration results of the training set after normalization, it can be found that PLU-MCN reaches convergence at about 90 rounds of iterations.

To demonstrate the superiority of the PLU-MCN model, 14 comparative models are selected in this paper, including ARIMA, LSTM, CNN [8], DCRNN, STGCN [12], ASTGCN, STSGCN [13], STFGNN, Auto-DSTSGN, STTN, GMAN, TFormer, ASTGNN, and PDFormer. Table 4.6 demonstrates that, with the evaluation metric of RMSE, the prediction performance of the PLU-MCN compared with the state-of-the-art models improved by 8.237%. This paper demonstrates the variation of RMSE accuracy for different models under PEMS03-G,

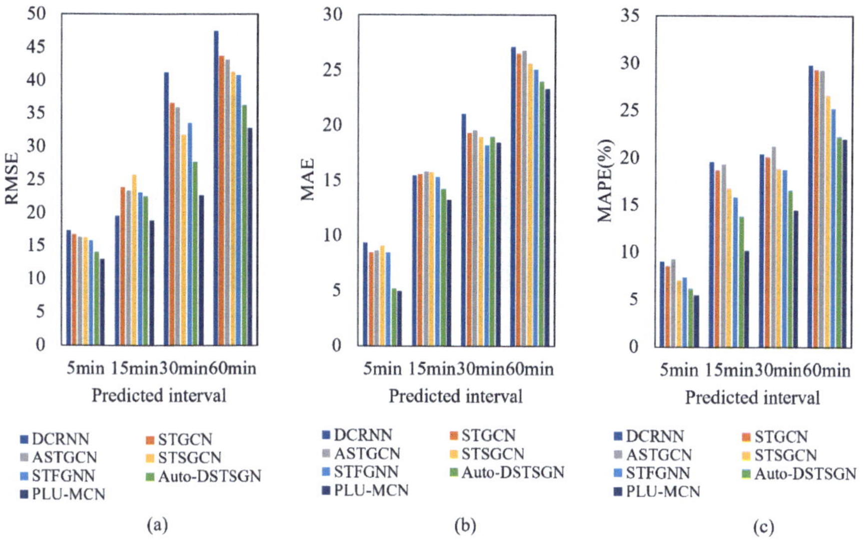

Fig. 4.23 Long-term predictive performance. (**a**) RMSE. (**b**) MAE. (**c**) MAPE (%)

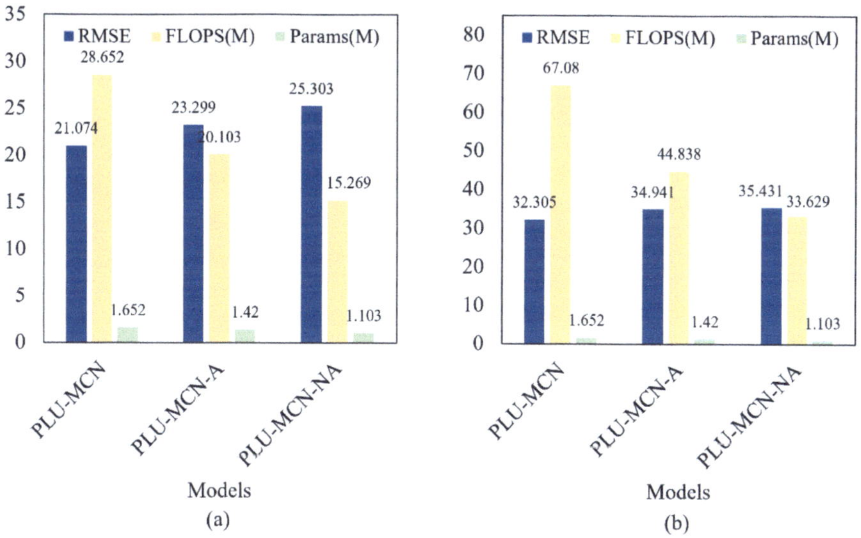

Fig. 4.24 Ablation experiment with node relationship matrix and attention mechanism. (**a**) PEMS03-G (gau = 0.2). (**b**) PEMS07-G (gau = 0.2)

PEMS04-G, PEMS07-G, and PEMS08-G datasets to verify the effect of perturbation data on prediction, as shown in Fig. 4.20. It can be found that with the increase of the Gaussian noise standard deviation gau, the PLU-MCN proposed in this paper is least affected by the noise and has good prediction performance.

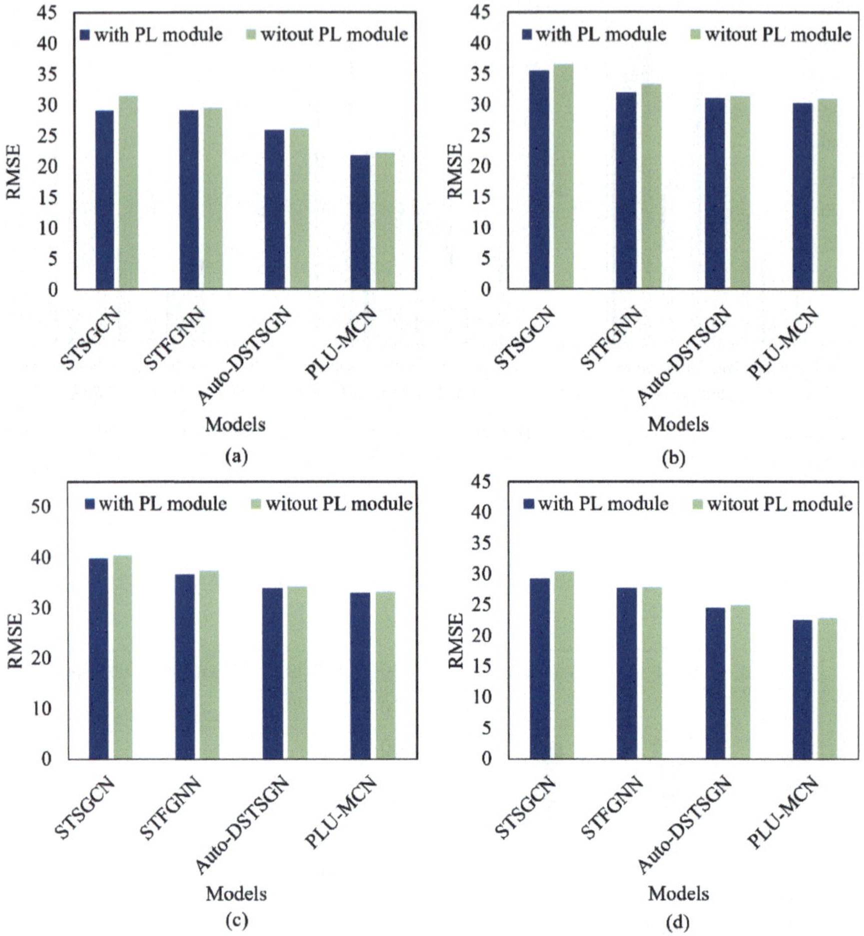

Fig. 4.25 Ablation experiment of PL module. (**a**) PEMS03-G (gau = 0.8). (**b**) PEMS04-G (gau = 0.8). (**c**) PEMS07-G (gau = 0.8). (**d**) PEMS08-G (gau = 0.8)

ARIMA is the classical time series prediction method, which is only applicable to the small sample simulation data. LSTM and CNN are the typical single-feature extraction models, which are difficult to mix the metrics of spatial and temporal features effectively. GCN- or attention-based models are more commonly used to extract spatial-temporal features of traffic flows, but they both ignore the impact of external factors and perturbation data on prediction performance.

To verify the performance of models in peak hour prediction, 8 a.m. to 10 a.m. and 5 p.m. to 7 p.m. of the SZ-TAXI and PEMS08 datasets are selected as two key time periods for testing in this paper, as shown in Fig. 4.21. PEMS04 is selected to test the prediction performance of PLU-MCN compared to the attention-based model for holidays, and as shown in Fig. 4.22, PLU-MCN has the optimal performance. The experimental results show that PLU-MCN has better prediction performance and can better perceive the traffic flow during peak hours. This paper divides the prediction period into 5, 15, 30, and 60 min to verify the long-term prediction performance of models as shown in Fig. 4.23. As the prediction period increases, the prediction performance of models decreases gradually, and the PLU-MCN proposed in this paper always has the optimal prediction accuracy. Figures 4.24 and 4.25 show the prediction performance of PLU-MCN under different variants of PLU-MCN, and PLU-MCN always maintains the optimal accuracy.

Although PLU-MCN exhibits excellent predictive performance on the five selected real datasets, it still needs to be optimized in terms of time complexity, training efficiency, and mathematical theory. In the future, we will focus on improving the model structure, increasing interpretability, and refining the mathematical derivation.

4.5 Conclusion

In summary, this chapter presents the advantages of hybrid models that integrate local and global spatial correlations for traffic prediction. By effectively combining different spatial feature extraction techniques, these models improve the understanding of traffic flow dynamics at various scales. The exploration of methods like the PLU-MCN and urban road network models showcases their ability to harness multi-level spatial information for enhanced prediction accuracy. This chapter concludes by reinforcing the necessity of developing hybrid approaches to achieve a comprehensive understanding of connected traffic systems, ultimately contributing to the advancement of intelligent transportation solutions.

References

1. N. Siddique, S. Paheding, C.P. Elkin, V. Devabhaktuni, U-Net and its variants for medical image segmentation: A review of theory and applications. IEEE Access. **9**, 82031–82057 (2021). https://doi.org/10.1109/ACCESS.2021.3086020
2. L. Zhao, Y. Song, C. Zhang, Y. Liu, P. Wang, T. Lin, M. Deng, H. Li, T-GCN: A temporal graph convolutional network for traffic prediction. IEEE Trans. Intell. Transp. Syst. **21**, 3848–3858 (2020). https://doi.org/10.1109/TITS.2019.2935152
3. C. Song, Y. Lin, S. Guo, H. Wan, Spatial-temporal synchronous graph convolutional networks: A new framework for spatial-temporal network data forecasting, in *AAAI(2020), AAAI, Palo Alto, USA*, (2020), pp. 914–921. https://doi.org/10.1609/aaai.v34i01.5438
4. C. Ding, J. Duan, Y. Zhang, X. Wu, G. Yu, Using an ARIMA-GARCH modeling approach to improve subway short-term ridership forecasting accounting for dynamic volatility. IEEE Trans. Intell. Transp. Syst. **19**, 1054–1064 (2018). https://doi.org/10.1109/TITS.2017.2711046
5. S. Hochreiter, J. Schmidhuber, Long short-term memory. Neural Comput. **9**, 1735–1780 (1997). https://doi.org/10.1162/neco.1997.9.8.1735
6. W. Zhang, Y. Yu, Y. Qi, F. Shu, Y. Wang, Short-term traffic flow prediction based on spatio-temporal analysis and CNN deep learning. Transportmetrica A: Transport Science **15**, 1688–1711 (2019). https://doi.org/10.1080/23249935.2019.1637966

7. Y. Li, R. Yu, C. Shahabi, Y. Liu, Diffusion convolutional recurrent neural network: Data-driven traffic forecasting, in *ICLR(2018), Vancouver, Canada*, (2018), pp. 1–16. https://doi.org/10.48550/arXiv.1707.01926
8. B. Yu, H. Yin, Z. Zhu, Spatio-temporal graph convolutional networks: A deep learning framework for traffic forecasting, in *IJCAI(2018), IJCAI, Stockholm, Sweden*, (2018), pp. 3634–3640. https://doi.org/10.24963/ijcai.2018/505
9. S. Guo, Y. Lin, N. Feng, C. Song, H. Wan, Attention based spatial-temporal graph convolutional networks for traffic flow forecasting, in *AAAI(2019), AAAI, Hawaii, USA*, (2019), pp. 922–929. https://doi.org/10.1609/aaai.v33i01.3301922
10. M. Li, Z. Zhu, Spatial-temporal fusion graph neural networks for traffic flow forecasting, in *AAAI(2021)*, (2021), pp. 4189–4196. https://doi.org/10.1609/aaai.v35i5.16542
11. G. Jin, F. Li, J. Zhang, M. Wang, J. Huang, Automated dilated spatio-temporal synchronous graph modeling for traffic prediction. IEEE Trans. Intell. Transp. Syst., 1–11 (2022). https://doi.org/10.1109/TITS.2022.3195232
12. M. Xu, W. Dai, C. Liu, X. Gao, W. Lin, G.-J. Qi, H. Xiong, Spatial-temporal transformer networks for traffic flow forecasting. ArXiv Preprint., 1–12 (2020). https://doi.org/10.48550/arXiv.2001.02908
13. C. Zheng, X. Fan, C. Wang, J. Qi, GMAN: A graph multi-attention network for traffic prediction, in *AAAI(2020), Palo Alto, USA*, (2020), pp. 1234–1241. https://doi.org/10.1609/aaai.v34i01.5477
14. H. Yan, X. Ma, Z. Pu, Learning dynamic and hierarchical traffic spatiotemporal features with transformer. IEEE Trans. Intell. Transp. Syst. **23**, 22386–22399 (2022). https://doi.org/10.1109/TITS.2021.3102983
15. S. Guo, Y. Lin, H. Wan, X. Li, G. Cong, Learning dynamics and heterogeneity of spatial-temporal graph data for traffic forecasting. IEEE Trans. Knowl. Data Eng. **34**, 5415–5428 (2022). https://doi.org/10.1109/TKDE.2021.3056502
16. J. Jiang, C. Han, W.X. Zhao, J. Wang, PDFormer: Propagation delay-aware dynamic long-range transformer for traffic flow prediction, in *AAAI(2023), Washington, USA*, (2023), pp. 1–9. https://doi.org/10.1609/aaai.v37i4.25556

Chapter 5
Summary and Future Challenges

5.1 Future Challenges

Although graph neural networks have achieved advanced performance in traffic prediction, there still exist some challenges that have not been fully addressed. Based on these challenges, potential future research directions have been proposed. These challenges and opportunities can be summarized as data heterogeneity, application diversity, multimodality, dynamism, model interpretability, uncertainty, and small sample learning.

Data heterogeneity: Traffic prediction data exhibits diversity in sources and complexity in structure and contains rich correlations. Some models use multiple graphs to encode different correlations, and constructing multiple graphs can help utilize traffic semantic information to improve prediction performance. Additionally, external factors also potentially influence traffic prediction outcomes, and incorporating diverse data sources such as POI data, event data, weather data, and social media data can also enhance prediction performance. Designing a framework for multi-source data fusion and extracting implicit domain knowledge from heterogeneous data are challenging. Hua et al. integrated multiple data sources including traffic data and weather data and added multi-layer knowledge graphs, while Zhu et al. constructed knowledge graphs and combined knowledge representation learning methods to integrate knowledge and traffic features as inputs for spatiotemporal graph networks. In the future, a knowledge graph can be used to intuitively describe the influence of external factors on traffic networks, integrate large-scale knowledge bases, optimize multi-source data structures, provide rich and diverse knowledge for traffic prediction, embed traffic knowledge into prediction models, and extract more semantic information.

Application diversity: Currently, there have been many studies using graph neural network architectures to solve traffic state prediction and demand prediction problems, achieving advanced prediction performance. However, there is less

Q. Shi et al., *Connected Vehicles Traffic Prediction*, Wireless Networks,
https://doi.org/10.1007/978-3-031-84548-2_5

research analyzing traffic data from a graph perspective in other traffic prediction tasks. Modeling other relational data as graphs expands the applicability of graph neural network models, such as modeling trajectory interactions. In complex pedestrian or vehicular roads, individual trajectory movement decisions are influenced by the interactions with other individuals. Considering coordinates, directional velocity, and inclinations as node features, introducing graph neural networks to encode interaction states and learn the impacts of interactions. Graph neural networks are also suitable for decision-making (traffic signal control) and detection tasks (anomaly detection), as well as tasks that exhibit data sparsity and high dimensionality, such as subway network-level OD prediction and individual travel destination prediction. Different tasks may have different graph structures, and extending graph neural networks to multi-task applications or integrating data from multiple tasks for training poses significant challenges. Future research can explore new application scenarios in the traffic domain from a graph perspective.

Multimodality: The transportation system consists of multiple modes, and to meet the diverse travel demands of passengers, coordination between different modes is essential for efficient transportation. Multimodal prediction has significant implications for managing and operating interconnected transportation systems. For example, accurate and timely multimodal prediction can facilitate intelligent deployment of ride-hailing and bike-sharing services in subway station areas, timely dispatching of connecting bus routes, and alleviation of sudden passenger flows in subways. One major challenge of graph-based multimodal prediction lies in the different spatial units of different transport modes. Some transport modes are based on stations, such as subways, while others are based on service areas, such as ride-hailing cars. Existing research has developed numerous GNN models based on homogeneous graphs composed of one type of node. However, they are not applicable to heterogeneous graphs consisting of multiple types of nodes. Recent studies have proposed heterogeneous graph embedding techniques, providing insights into solving this issue and applied in related fields such as link prediction and node classification. Due to the more complex and dynamically changing relationships between different modes, it is more challenging for the heterogeneous dynamic network of the transportation system. Liang et al. proposed a multi-relation spatiotemporal graph neural network prediction framework, which considered the heterogeneous spatiotemporal correlations among different spatial unit modes, differentiated the contributions of each relation to target nodes through attention-based aggregation modules, and explored the interactions between different modes. Future research can further analyze the significant influences between different modes, explore semantic associations between different modes, and apply transfer learning methods to improve prediction performance.

Dynamism: Both traffic networks and node features evolve over time and exhibit strong spatiotemporal correlations. Predefined matrices cannot capture the dynamism effectively, and modeling the evolution characteristics of dynamic graphs can help capture more effective information than prior spatial information. Modeling the dynamism of traffic networks presents significant challenges due to the complex spatiotemporal correlations involved. Future research can consider three aspects:

learning time-varying parameter matrices through data-driven approaches or generating dynamic matrices at each timestamp based on historical states and current information; introducing spatiotemporal graphs to encode complex spatiotemporal relationships from different timestamps, breaking the barriers between separate considerations of time and spatial correlations; constructing dynamic hyperedges to simulate dynamic spatiotemporal relationships, surpassing the limitations of pairwise relationships between nodes.

Model interpretability has become an important issue in the context of big data, particularly in the field of transportation operations. As the volume of data generated from transportation records continues to increase, traditional traffic flow methods struggle to handle such massive amounts of data. Deep learning techniques have emerged as a necessary trend for analyzing and predicting transportation big data. However, deep learning suffers from the drawback of being a "black box" model, lacking interpretability.

While graph neural network (GNN) models demonstrate promising predictive performance, they inherently lack interpretability. Developing GNN-based traffic prediction models that are easily interpretable presents a significant challenge. Graph structures contain additional relational information, which places higher demands on interpretability. One potential research direction is to integrate domain knowledge into data-driven networks, linking the parameters learned by GNN models with traditional traffic flow theories. By developing an integrated framework that combines model-driven and data-driven approaches, interpretability can be enhanced, providing support for transparent decision-making.

Currently, most GNN-based traffic prediction methods focus on improving the accuracy of prediction results. However, for high-risk applications such as automatic driving trajectory prediction, decision-makers not only need timely and accurate deterministic estimates of future traffic states but also require the model to precisely quantify prediction uncertainty. For example, allocating confidence values to prediction results would be ideal.

Small sample learning: The number of training samples required for traffic prediction models based on graph neural networks is proportional to the size of the traffic network. However, when the sample size is small, the performance of most models tends to significantly decline.

Currently, in an era where privacy protection is of utmost importance, it is challenging to meet the application-level requirements by training models using publicly available traffic datasets. Developing traffic prediction models with small-scale training data remains an unresolved issue.

5.2 Summary

The book provides a comprehensive exploration of the historical evolution of neural networks and graph neural networks within the context of traffic flow prediction for Connected Vehicles (CVs). By emphasizing the importance of traffic flow prediction in connected vehicle systems, it highlights its essential role in optimizing traffic management, enhancing road safety, and ensuring efficient resource allocation within intelligent transportation systems. The book also reviews a broad range of related research, offering readers valuable foundational knowledge on the topic.

Each chapter is carefully organized, with a dedicated focus on specific methods used in traffic flow prediction. These chapters delve into problem definitions, model architectures, experimental setups, and result discussions, offering in-depth insights into each methodology. This structure makes the book a valuable resource for both researchers and practitioners in the field of traffic flow prediction for connected vehicles.

A notable feature of the book is its effective application of deep learning techniques, particularly graph neural networks, for traffic flow prediction. These techniques have shown significant potential in addressing the unique challenges posed by CV systems and provide promising avenues for advancing intelligent transportation.

As technology advances, the informatization of traffic systems within the connected vehicle landscape is expected to accelerate, facilitating interdisciplinary integration with other technological domains. In line with this trend, the book offers several recommendations for future research in CV-related traffic flow prediction:

1. Optimizing model efficiency: Although many graph neural network models focus on enhancing prediction accuracy, they often increase training complexity. Given the large-scale traffic networks in connected vehicle systems that demand substantial computational resources, it is crucial to optimize model structures or develop lightweight models that maintain performance while aligning with practical applications.
2. Incorporating emerging technologies: Future research should explore the integration of emerging technologies like reinforcement learning, transfer learning, and federated learning within CV frameworks. These technologies expand the range of applications, address challenges such as limited sample sizes and data privacy, and support the continued advancement of CV-based traffic prediction.
3. Enhancing model interpretability: Future research should emphasize model interpretability and investigate the synergy between model-driven and data-driven approaches. This could involve combining traditional traffic flow theories, domain expertise, and data inference methods, enabling a more profound understanding of the mechanisms behind traffic flow prediction in connected vehicle systems.

In essence, the book serves as a valuable guide, shedding light on the historical development, current trends, and future directions of graph neural network-based approaches to traffic flow prediction in connected vehicles. The suggested directions for future research are poised to drive ongoing progress in the field, ultimately advancing the capabilities of intelligent transportation systems for a more connected and efficient future.

Index

Q. Shi et al., *Connected Vehicles Traffic Prediction*, Wireless Networks, https://doi.org/10.1007/978-3-031-84548-2

The manufacturer's authorised representative in the EU is Springer Nature Customer Service Centre GmbH, Europaplatz 3, 69115 Heidelberg, Germany. If you have any concerns regarding our products, please contact ProductSafety@springernature.com

Printed and bound by CPI Group (UK) Ltd, Croydon, CR0 4YY
30/09/2026
02225598-0001